BASIC CARTOGRAPHY
for students and technicians

Volume 1

2nd Edition

BASIC CARTOGRAPHY
for students and technicians

Volume 1

2nd Edition

Editors

R. W. ANSON

and

F. J. ORMELING

Published on behalf of the
INTERNATIONAL CARTOGRAPHIC ASSOCIATION
by
ELSEVIER APPLIED SCIENCE PUBLISHERS
LONDON and NEW YORK

Distributed by
PERGAMON PRESS Ltd
Headington Hill Hall, Oxford, OX3 0BW, UK

WITH ILLUSTRATIONS

© 1993 ELSEVIER SCIENCE PUBLISHERS LTD

British Library Cataloguing in Publication Data

Library of Congress Cataloging-in-Publication Data
Basic cartography for students and technicians / editor, R. W. Anson
    and F. J. Ormeling. -- 2nd ed.
        p.   cm.
    Includes index
    1. Cartography.   I. Anson, R. W.   II. Ormeling, F. J.
    III. International Cartographic Association.
GA105.3.B38   1993
526--dc20   93–35427   CIP
ISBN 0-08-042343-4 Hardback
        0-08-042344-2 Paperback

Photoset in Great Britain by BPCC Techset Ltd, Exeter.
Printed and bound in Great Britain by
Butler & Tanner Ltd, Frome and London

CONTENTS

FOREWORD

The preparation of a text relating to *Basic Cartography for students and technicians* was first suggested at the International Cartographic Association's 6th Conference which was held in Montreal and Ottawa, Canada, in 1972. It was widely felt that there was a requirement for a well-illustrated, instructional book on cartography written in reasonably simple language.

Originally conceived as a single tome, the multi-author work was subsequently divided into two separate but complementary volumes in order to expedite production and reduce costs. The first book was launched in 1984 at the 12th International Cartographic Conference which was held in Perth, Australia, and the second in Budapest, Hungary, during the ICA's 14th World Conference in 1989. In 1991 these were supplemented by an *Exercise Manual*, the contents of which were directly linked to the chapters contained in the two volumes. All three publications have enjoyed considerable success, and ICA owes a debt of gratitude to the international team of specialist authors and editors who worked on the trilogy. Thanks are also due to the numerous organisations, in ICA member countries throughout the world, who very generously provided assistance of many kinds.

The English-language version of Volume 1 has been translated by colleagues in China, Mexico and Thailand to enhance its usefulness in their respective countries. It is now out-of-print, and the opportunity has been taken to revise, re-edit and update the contents to provide a better reflection of cartographic procedures in the 1990s.

The ICA wishes to record its sincere thanks to all contributors for their assistance in preparing the new English edition of this important contribution to the literature available for use in cartographic education. Its publication demonstrates the ongoing involvement and interest of the ICA's Commission on Education and Training in this field.

D. R. F. Taylor
President, International Cartographic Association

PREFACE

Maps have been produced and used for some 5 000 years, the earliest extant examples dating back to the third millennium BC, but the significance and importance of these media of graphic communication has only begun to be generally appreciated comparatively recently. The term 'cartography' did not become current until 1839, and the first internationally recognised definition of the subject was not published until 1973. At that time it was stated that 'Cartography is the art, science and technology of making maps, together with their study as scientific documents and works of art. In this context maps may be regarded as including all sorts of maps, plans, charts and sections, three-dimensional models and globes representing the Earth or any celestial body at any scale.'

Almost contemporaneously with the issuing of this explanation, the International Cartographic Association (ICA) assembled a team of academic and professional cartographers in an attempt to close a perceived gap in available educational literature. This was planned to be done by producing a work, suitable for use worldwide, written in simple language and employing a highly graphic approach to enhance readers' understanding of technical details incorporated in the text. Further, it was envisaged that the adopted scheme would facilitate translation and republication in languages other than that of the orginal English version. Volume 1 of *Basic Cartography for students and technicians* was launched during the 12th International Cartographic Conference which was held in Perth, Australia, in 1984. Subsequently, it has been translated to enable its use in China, Mexico and Thailand.

Much has happened since the work was published and cartography, like many other facets of modern life, has become increasingly embroiled in the throes of a revolution involving the application of electronic-based information technology and the output of computer-generated products. The employment of digital databases, and the utilisation of Geographic Information Systems, are becoming evermore common worldwide with a resultant modification of the scope of cartography and the type of products available. Of necessity, the term required redefinition to reflect these new approaches and a special working group of the ICA has now proposed that cartography is 'the discipline dealing with the conception, production, dissemination and study of maps'.

The first edition of Volume 1 of *Basic Cartography* now being out-of-print, the opportunity has been taken to subject the original content to a complete revision, re-edit and update in order to take into account the recent developments relating to the subject and the practices employed in the 1990s. The initial objectives of the project have again been adhered to in this second edition, but authors have modified their contributions to recognise new concepts and developments of relevance to both students and technicians.

Special thanks are due to the individually named authors who have prepared chapters for inclusion in this work, and also to members of the French Commission on Education in Cartography and the Working Group of the Japan Cartographers Association who contributed to Chapters 3 and 4 respectively. The assistance and encouragement provided by members of the ICA Executive Committee is gratefully acknowledged, as is also the support provided by its Commission on Education and Training and Publications Committee.

R. W. Anson
F. J. Ormeling

THE AUTHORS

Richard E. Dahlberg is the USA's representative member and co-chair of the ICA's Commission on Education and Training for the period 1991–1995. He is Professor of Cartography in the Department of Geography at Northern Illinois University, and also Director of its Laboratory for Cartography and Spatial Analysis. During 1988–1989 he was editor of the *American Cartographer*, and in 1990 *Cartography and Geographic Information Systems*. He has been particularly involved in developing the potential applications of appropriate technology and the interface between remote sensing and cartography. Currently he is President of the American Congress on Surveying and Mapping.

Prof. Dr W. C. Koeman is Emeritus Professor of Cartography at the University of Utrecht. The Netherlands, and was Chairman of the ICA's Commission on Education in Cartography from 1972 until 1980. He started his career as a cartographic draftsman, and then studied Geodesy at the Technological University in Delft. In 1961 he prepared his doctoral thesis on the 'Collection of Maps and Atlases in The Netherlands', and in 1967 became his country's first holder of a chair in cartography. Professor Koeman is the author of many publications relating particularly to the history of cartography.

Dr D. H. Maling originally served in the Royal Air Force as a navigator, but was then retrained as a meteorologist and worked in this capacity for two years in Antarctica. From 1955 until 1980 he taught cartography in the Department of Geography at the University College of Swansea (UK). He is an internationally known specialist on map projections and the first edition of his book *Coordinate Systems and Map Projections* was published in 1973. In 1989 this was followed by *Measurements from Maps*, and he has also produced numerous research articles on this subject. Dr Maling has been a member of many national and international committees, and was formerly the UK's representative on the ICA's Commission on Education in Cartography.

Dr B. Rouleau initially studied geography and then entered the Ecole Supérieure de Cartographie Géographique, University of Paris, which he has directed since 1962. He is Maître Assistant and organises courses in cartography at all levels. Since 1973 he has been Chairman of the French Commission on Cartographic Education, and has been his country's representative on the ICA's Commission on this topic since 1966. His special interests are in urban mapping and the history and growth of towns, and his book *Le Tracé des Rues de Paris* (CNRS) has been reprinted several times. He prepared his thesis for the 'Doctorat d'Etat' (Sorbonne, 1982) on the evolution of the suburbs of Paris (*L'Espace Urbain Parisien à Travers ses Cartes*), and this was published in 1984.

Kei Kanazawa was employed as a professor at the Kochi Technical College (Japan) from 1979 until 1984, and has been a member of the ICA's Commission on Cartographic Education and Training since 1964. He started his career as a draftsman, but then studied cartography and surveying at the Construction College, Ministry of Construction, and differential geometry at the Tokyo Municipal University. Having obtained a Master's degree in mathematics (map projection theory) in 1962, he registered as a Government Certified Consultant Engineer (applied science) in 1976. He is currently senior adviser to the Tokyo Cartographic Co., and his numerous publications show particular emphasis on cartographic design techniques.

Christer Palm began his career in 1946 as an army officer serving at the Swedish Geographical Survey Office, and from 1947 until 1960 worked there as a civil servant. He then joined Esselte Map Service (1960–1972), and subsequently the Hydrographic Department (1972–1982) where he was head of the charts section. In addition, he has been both Secretary (1966–1971) and Chairman (1971–1974) of the Cartographic Society of Sweden, and his country's representative on the ICA's Commission on Education and Training. Since 1968 he has taught cartography at the Swedish Royal Institute of Technology in Stockholm, and is a former member of the board of the National Orienteering Association with special responsibility for map production. He has published many papers on a wide variety of cartographic topics.

Sjef van der Steen works in the Cartography Division of the International Institute for Aerospace Survey and Earth Sciences (ITC), Enschede, The Netherlands. In 1967 he began studies as an intaglio photographer/retoucher and gained the Diploma; in 1972 he commenced training to become a Practical Teacher of Graphic Arts. He joined ITC in 1974 and has worked in both the Aerial Photography and Cartography Divisions. At present he is a full-time instructor in repro-photography, and teaches both pre-press production management and manual and automated map generation. Externally he is a member of the Dutch Cartographic Society's Commission on Map Production Techniques, and the ICA's Commission on Map Production Technology.

THE EDITORS

Roger W. Anson is the Principal Lecturer in Cartography at Oxford Brookes University (UK). After graduating from the University of Oxford with an Honours degree in Geography he worked in the cartographic department of Pergamon Press Ltd from 1965 to 1970, eventually becoming Chief Cartographic Editor. Currently he is the Vice President of the British Cartographic Society and Chairman of its Executive Committee; a member of the UK Committee for Cartography; and his country's representative on the ICA's Commission on Education and Training. In addition, he has been Chairman of the ICA's Publications Committee since 1981, and is co-editor of the ICA Newsletter. Contributions were made to an ICA–CET seminar in Bangkok (Thailand), and personal publications include atlases, sheet maps, and a significant number of papers and contributions to texts relating to a variety of aspects of mapping.

Ferjan Ormeling was trained as a geo-cartographer and has, since 1973, been The Netherlands' representative on the ICA's Standing Commission on Education and Training of which he is at present co-chair. Further, he is a member of the ICA's Commission on National and Regional Atlases; Vice President of the Dutch Cartographic Society; and Professor of Cartography at the University of Utrecht. Before moving to Utrecht in 1969, he worked for eight years with Wolters-Nordoff Atlas Productions in Groningen. He is a member of the editorial board of the National Atlas of The Netherlands, and chief editor of *Kartografisch Tijdschrift* (the journal of The Netherlands' Cartographic Society). Research interests include thematic mapping and toponomy, and he has participated in ICA–CET seminars in Rabat (Morocco), Wuhan (China) and Bangkok (Thailand).

INTRODUCTION

R. E. Dahlberg

Preamble

These pages serve to introduce the subject of cartography as portrayed by authors contributing to *Basic Cartography for students and technicians*, and attempt to characterise the institutional and disciplinary milieu in which the work was created. Its conception and publication is evidence of the great importance that the International Cartographic Association (ICA) attaches to education and training in the field of cartography. This commitment is further reflected in the programmes of ICA conferences and seminars, in its publications, and in its Commission structure. The original Commission on Education was one of the first three interest groups established as long ago as 1964, the year during which the ICA was initially exposed to the potential afforded by adoption of an automated cartographic system at its conference in London (*ICA 1959–1984*, F. J. Ormeling, Snr, 1987). Of necessity, the early years of the Commission's activities were focused upon the basic tasks of nurturing awareness and interest in education at ICA conferences, and taking stock of the structure and resource base of cartographic instruction at a global level.

In 1972 the chairmanship of the Commission passed to Professor C. Koeman (Netherlands) who initiated the development of a multilingual cartographic manual aimed at those working at technician level. The plan envisaged was the preparation of a volume relating to mainstream cartography, and consisting of approximately equal proportions of text and illustrations. The degree of emphasis upon the latter was intended to facilitate the subsequent generation of editions in languages other than the two used officially by the ICA (English and French). The scheme proved to be both complex and difficult, and work on the manual progressed very slowly. In 1981 the project was transferred to the ICA's Publications Committee, which decided to bring together the chapters completed at that time and to issue them as Volume 1. This appeared in 1984, and the remaining chapters of the originally conceived work were published in 1988 as Volume 2.

During the 1980s the objectives of the renamed Commission on Education and Training consisted of the accomplishment of two major tasks. The first of these was the preparation of an Exercise Manual to accompany the two volumes of *Basic Cartography* (*The Exercise Manual*, F. J. Ormeling, Jnr, 1989). Produced with minimal financial assistance from the ICA, this proved to be an extraordinarily complex project, but it was finally completed and issued in 1991. The work consists of more than seventy exercises based on the content of the two volumes. A second task was the formulation of a seminar programme dedicated to the study of cartography, and focused on the immense tasks relating to technology transfer (*An ICA Response to the Educational Challenge of Cartography in Transition*, R. E. Dahlberg, 1987). The problems involved in conceptualising the content and emphases of seminar contributions were both challenging and difficult. As a result, the Commission became involved in the tracking of significant shifts in the approach to and development of the subject and its related sister disciplines. However, at all times members maintained their interests in and concerns for mainstream cartography. As a result of experience gained from the planning and delivery of its seminars, the Commission was better prepared to address the task of preparing a second edition of *Basic Cartography*, Volume 1. In addition, it is undertaking the organisation and compilation of a third book in the series, relating to the investigation and application of new developments applicable to the field. It is heartening to report that the Commission's resolve to maintain the currency and relevance of *Basic Cartography* continues to enjoy strong support from the ICA leadership, and that the policy of facilitating its translation into other languages is being actively pursued.

Volume 1

This second edition of Volume 1 preserves the character and broad international base of authorship employed for the original publication. Further, the emphasis on explanation through graphics has been retained. Considered together, Volumes 1 and 2 provide an authoritative and comprehensive view of the subject as seen during the early stages of the transition from a conventional analogue, or graphics-based, discipline to a database technology. The intention in producing a third volume is to complement the two earlier works with materials providing a current view of a more mature, digital technological base.

In the first chapter, Professor C. Koeman presents selected highlights of the **History of Cartography** spanning six millennia. Drawn from a diversity of cultures and eras, the examples cited and illustrated relate to mapping produced in Mesopotamia, Egypt, China, Japan and the Moslem world as well as Western Europe. Major changes in styles, composition and types of maps, together with the methods employed in their construction, are fully described. Despite its brevity, the contribution conveys the notion of the enduring and universal importance of maps, and of a rich cartographic heritage worthy of further investigation and study.

In Chapter 2, Dr D. H. Maling provides a lengthy and richly illustrated exposition with regard to **Mathematical Cartography**. The basic terminology relating to scale, grid systems, and map format is explained, and important concepts concerning the shape and size of the Earth, geographical and plane co-ordinate systems are detailed. An especially thorough discussion is provided of the methods of plotting, hand-drafting, or scribing the geometric framework of a map. The rest of the chapter is devoted to an examination of map projections with respect to background theory, fundamental and special properties, and the main classes of graticules. The distributions of scale variations and angular deformation are described by both numerical and graphical means, including the use of distortion ellipses.

This is followed by a chapter on the **Theory of Cartographic Expression and Design**, which has been prepared by Dr B. Rouleau in collaboration with numerous French colleagues. It is a delightful meld of geography, graphic language and thematic cartography which complements the contributions on Map Compilation and Thematic Cartography contained in Volume 2. The profusely illustrated text poses questions of fundamental importance such as 'What purpose does a map serve?' and 'What sort of a map should be made?'. The discussion of data includes sources of information; the meaningful matching of graphic variables to the levels of organisation of information components; and ways of achieving visual synthesis. A section on graphic representation considers symbol design, size, texture and structure, value, grain, colour, orientation and shape. A description of the rules of graphic language focuses on the relationship between visual variables and symbol form and levels of measurement, and clearly illustrates the topics of perception, separation and differentiation. The chapter concludes with a concise explanation of traditional representation models which provide for the appropriate pairing of symbol form and distribution type.

Writing on **Techniques of Map Drawing and Lettering**, Professor K. Kanazawa, together with a sizeable working group, provides a detailed and highly graphic account of conventional instruments, tools and materials, pen and ink drafting; scribing; masking and stick-up; and lettering. Users of this chapter will particularly appreciate the eight pages devoted to the high-quality illustrations of hachures, rock symbolisation and hill-shading. The many paired comparisons demonstrating the 'right' and 'wrong' ways of achieving a desired effect incorporate smiling or frowning facial icons.

The volume is concluded by a chapter entitled **Cartographic Pre-Press, Press and Post-Press Production**, which has been compiled by Messrs. C. Palm and S. van der Steen. Like the previous contributions it is lavishly illustrated, and succeeds in providing a concise explanation of the essentials of the map reproduction processes and the equipment required to implement them.

Volume 2

As was mentioned earlier, the publication in 1988 of Volume 2 concluded the original scheme, formulated in 1972, for the production of a manual devoted to basic cartography. Subsequently, the 'opus' has been supplemented by an *Exercise Manual*, and the intended issuing of a third volume will round-off and balance the mix of conventional and digital cartography. Here, as in Volume 1, a generous amount of space has been devoted to explanatory graphics.

In the opening chapter, Dipl. Ing. R. Böhme succeeds in describing the vast field of **Topographic Cartography** within a modest space. The discussion includes reference to different scale classes, together with the representation of relief, natural, and man-made features. Examples of detail depiction are provided for mapping at scales varying from 1 : 5 000 to 1 : 500 000. A brief but useful explanation of generalisation is accompanied by eight map extracts at scales between 1 : 5 000 and 1 : 200 000. The importance of geographical nomenclature and the standardisation of names is also stressed, and map revision and the significance of maintenance programmes are discussed. The chapter concludes with comparative statistics relating to the percentage coverage of continents by topographic maps at 1 : 25 000. These data remind us that, worldwide, much primary mapping still remains to be completed. Numerous basic references are also provided in the attendant bibliography.

Chapter 2 constitutes approximately one-third of this volume, and consists of thoughts on **Map Compilation** documented by Professor E. Spiess. It is generously provided with illustrations which are well integrated with the concise text. There are numerous examples of 'good' and 'poor' practice, each marked by smiling or frowning facial icons. Some readers may be surprised to find so much design-related content in a chapter devoted to the compilation process, but topics explored are illustrated by very relevant, realistic and high-quality examples. Two aspects pertaining to map design are worthy of special mention here. Firstly, the description of area colour selection which features some especially effective illustrations, and secondly, the challenging process of creating an appropriate visual hierarchy.

Generalisation, or cartographic abstraction, the task or process of transforming geographical reality into a mapped representation, is described by Mr M. J. Balodis in Chapter 3. Many aspects of the complex modification process are discussed and include selection, classification, simplification, exaggeration and symbolisation. Two basic considerations that influence and direct the generalisation process are map purpose and the intended user. In the case of topographic mapping, rules governing the processes have evolved over a considerable time and through practice, and they are understood and appreciated by experienced users. However, in thematic representations the wide diversity of topics, purposes and users has led to more variation in the use of generalisation rules—a fact not often evident to a user. Of special interest are examples showing how generalisation influences the abilities of map users to make logical inferences from a product.

In Chapter 4, **Thematic Cartography** is presented by its authors, Professor R. Ogrissek and Ing. E. Lehmann, as a subdiscipline of the subject, which is playing an increasingly important role in fields such as planning, environmental management and diverse scientific disciplines. Additionally, thematic cartography is considered in the context of a developed map production structure in which the responsibilities of participants tend to be formalised. In adopting this policy, the authors have drawn upon a rich experience base to create a window into the mature realm of thematic cartography which offers many lessons to professionals and students at all levels. Numerous comparisons are drawn between general cartography, and its products, and the more specialised field of thematic mapping. Emphasis is particularly devoted to the communication function of thematic products designed for use by specific audiences. The main types of maps are described, and the varieties of symbols employed are lavishly illustrated. The need is stressed for an editorial plan to document all stages in the preparation of a map, and also the importance of generalisation and standard specifications. A discussion of thematic map design reinforces and complements those appearing in Chapter 3 of Volume 1, and Chapter 2 of Volume 2. The importance of carefully preparing legend details

for most thematic maps is also strongly advocated.

The final chapter, written by Professor N. Kadmon, provides an introductory treatment of four aspects of **Computer-Assisted Cartography**. Firstly, there is a general description of digital computers, which features many illustrations; secondly, the hardware used for data capture and output is explained; thirdly, applications are reviewed with respect to editing, change of scale, rotation, map projections, digital map lettering, etc.

The final aspect considered is the operation of basic equipment.

Volume 3

In order to understand the niche that Volume 3 is intended to fill it is necessary to appreciate how the field of cartography was viewed in the early post-World War II years. The United Nations adopted the following definition of cartography in 1949, in order to provide a frame-of-reference for its operations:

> Cartography is considered as the science of preparing all types of maps and charts, and includes every operation from original survey to final printing of maps (United Nations, 1949).

When the International Cartographic Association was established in 1959, one of its earliest and most important concerns was how it should delineate the scope of the discipline. In his study entitled *ICA 1959–1984: The First Twenty-Five Years of the International Cartographic Association*, Professor F. J. Ormeling, Snr, gives the following helpful insight on the ICA perspective:

> Contrary to the broad UN concept of cartography the Committee of Six concentrated on a more restricted field, excluding surveying and photogrammetry and all primary data gathering by other disciplines such as geology, statistics, etc. (ICA, 1987).

At its meeting in Amsterdam in 1967, ICA adopted the following definition of cartography, which subsequently appeared in the *Multilingual Dictionary of Technical Terms in Cartography*:

> The art, science and technology of making maps, together with their study as scientific documents and works of art. In this context maps may be regarded as including all types of maps, plans, charts and sections, three-dimensional models and globes representing the Earth or any celestial body at any scale. (ICA, 1973).

Since the formulation and adoption of the ICA definition, remarkable developments in technology, together with the so-called Information Revolution, have led to a widely felt need for a new and broader definition. Following several years of study and discussion, an ICA Working Group on Cartographic Definitions has reported that general agreement has been reached on the following:

> **Map:** A conventionalised image representing selected features or characteristics of geographical reality

designed for use when spatial relationships are of primary relevance.

Cartography: The discipline dealing with the conception, production, dissemination and study of maps.

Cartographer: A person who engages in Cartography.
(Dr C. Board, 1991)

In their conceptualisation and preparation of Volume 3, members of the Commission on Education and Training have significantly broadened the scope of the subject's field as it was embodied in Volumes 1 and 2 of *Basic Cartography*. In responding to a need to incorporate detail relating to significant developments in information technology, and to changing societal demands for spatial data, the Commission has, nevertheless, attempted to adhere to its original mandate. Thus the focus is still upon **basic** cartography and its current scope. The third volume, at present in the process of preparation, will contain contributions on the following topics: Communication, Design and Visualisation; Map Revision—traditional and digital modes; Technical Aspects of Remote Sensing; Cartographic Applications of Remote Sensing; Toponomy; Geographic Information Systems and Cartography; and Desk-Top Cartography.

REFERENCES

Board, C., 1991. Defining what we do, *Cartographic Perspectives*, **11** (Fall), 25–26.

Dahlberg, R. E., 1987. An ICA response to the educational challenge of cartography in transition, *Cartographica*, **24**, 1–13.

International Cartographic Association, 1973. *Multilingual Dictionary of Technical Terms in Cartography* (Ed. E. Meynen).

International Cartographic Association, 1987. *ICA 1959–1984: The First Twenty-Five Years of the International Cartographic Association*, compiled by F. J. Ormeling, Snr.

International Cartographic Association, 1988. *Basic Cartography for students and technicians*, Vol. 2 (Ed. R. W. Anson). ICA and Elsevier Applied Science Publishers.

International Cartographic Association, 1991. *Basic Cartography for students and technicians, Exercise Manual* (Eds R. W. Anson and F. J. Ormeling). ICA and Elsevier Applied Science Publishers.

International Cartographic Association, Commission on Education and Training. *Proceedings of the Seminar on Teaching Computer-Assisted Map Design*, held at Siemens AG, Munich, May 9–11, 1988.

International Cartographic Association, Commission on Education and Training. *Proceedings of the Seminar on Teaching Cartography for Environmental Information Management*, held at ITC, Enschede, April 24–26, 1989.

International Cartographic Association, Commission on Education and Training. *Proceedings of the Seminar on Teaching the Interface between Cartography, Remote Sensing and GIS*, held at the Eötvös Lorand University, Budapest, August 15–16, 1989.

International Cartographic Association, Commission on the History of Cartography and the Commission on Education and Training. *Proceedings of the Seminar on Teaching the History of Cartography*, held at the University Library, Uppsala, June 23, 1991.

Ormeling, F. J., 1989. The *Exercise Manual*: A companion to *Basic Cartography*, paper presented at the ICA 14th World Conference, August 17–24, Budapest.

Taylor, D. R. F., 1991. A conceptual basis for cartography/new directions for the information era, *Cartographica*, **28** (4) (Winter), 1–8.

United Nations, 1949. *Modern Cartography, Base Maps for World Needs*, p. 7. Publ. No. 1949. I. 19.

Chapter 1

THE HISTORY OF CARTOGRAPHY

C. Koeman

CONTENTS

1.1   Background

A visitor to the offices of an atlas publisher remarked to one of the editorial staff 'I suppose you go up with a camera to photograph countries for your maps?'

Professional cartographers will be amused by this misconception of their work, but it is a fact that the majority of people have not the remotest idea as to how maps are made! However, in one respect both professionals and non-initiates are in agreement in that they are all aware that modern maps are not merely the endproducts of imagination. When similar parameters are employed in considering maps prepared many centuries ago, the same is also generally true. Examples of mapping generated by civilisations which flourished thousands of years earlier demonstrate that the peoples of these times were also environmentally aware. No matter how far one looks back it is apparent that there have always been those capable of producing reasonably accurate portrayals of the distribution of spatial features from memory. However, these 'mental maps' were not based on formally organised surveys because this science was not sufficiently developed to allow the precise measurement of significant dimensions. Although people appreciated the relative spacing of geographical phenomena, as a result of their general environmental awareness, the distance between them was measured merely by comparing travelling times which were assessed with reference to the variations in position of the sun during daylight hours.

In Europe the use of the magnetic compass was unknown until c.1300, but here again, prior to its introduction, the sun served as an aid to orientation and the determination of direction. The beginnings of the sciences of surveying and navigation also mark the start of the history of cartography, for without accurate observations relating to distance and direction, effective mapmaking is not feasible. Nevertheless, during some periods of time maps were drawn based purely on imagination, with men believing the shape and extent of the world to be whatever they considered appropriate and convenient to them. Examples are provided by presumptions concerning a small, habitable world made by Greek philosophers in about 600 BC; and by the concept of legendary islands thought to lie in the Western Atlantic as recently as 1400 AD. These imaginary ideas relating to planet Earth, or to the Universe, were not conceived by practical surveyors but rather by philosophers, theologians and similar theorists. Mankind is deeply indebted to astronomers and courageous navigators, explorers and travellers for the eventual elimination of misconceptions concerning the nature of our world.

The science of navigation was introduced into Europe at rather a late stage as compared with the history and development of the subject in China. Sundials, levelling instruments and compasses were regularly mentioned in Chinese technical writings dating from c. 1100 BC. Western historians are inclined to overlook the knowledge gained as a result of studying non-European map sources during the Great Age of Exploration and

Fig. 1.1(a) Eskimo map of Kronprinsens Ejland near Godhavn, Greenland

Fig. 1.1(b) Aerial photograph (scale c. 1 : 60 000) of Kronprinsens Ejland for comparison with Fig. 1.1(a)
(Geodetic Institute, Copenhagen)

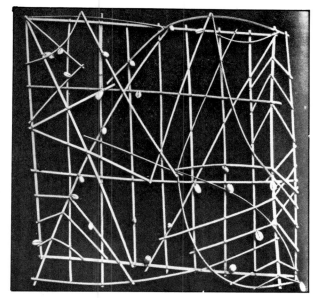

Fig. 1.2(a) 19th-Century navigational chart produced by Marshall Islanders from sticks and shells (British Museum, London). See: Lyons, H., The sailing charts of the Marshall Islanders, *Geographical Journal*, **lxxii** (4), October 1928

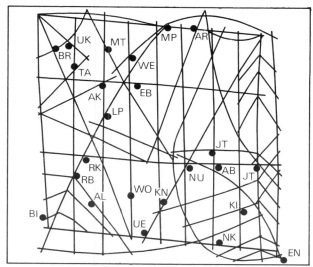

Fig. 1.2(b) Graphic reconstruction of Fig. 1.2(a)

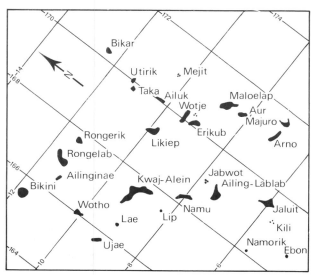

Fig. 1.2(c) Modern map of the same area

Discovery. For example, in 1498 Vasco da Gama, the Portuguese discoverer of the sea route to India, hired an Arab pilot who possessed a written pilot book and a chart of the coasts around the Indian Ocean. In about 1520 the Spanish 'conquistadores', under Cortez, looted their way through Mexico guided by very reliable Aztec maps. Unfortunately only two copies of Aztec maps compiled prior to the arrival of the Spanish have survived, because subsequently Catholic priests systematically burned all that they considered to be heathen documents. In due course, with the exception of China, Japan, Korea and Russia, the whole world was surveyed and mapped by the geodesists and cartographers of colonising nations. Initially, the surveys assisted merely in the exploration and exploitation of the overseas territories, and investigations were pursued only as far as was deemed necessary for profit making. Thus vast regions in the interiors of the continents of Africa and South America, for example, remained unmapped because the colonial powers saw no immediate financial gain resulting from such an exercise. This situation effectively persisted until the 1950s when a number of new, independent states were created and others began to seek self-government.

Textbooks on the history of cartography frequently document developments as viewed from the individual perspectives of their respective authors, and often reflect a specialist writer's considerable knowledge of mapmaking in his own country. However, in an international text of this type a chapter on the history of the subject should be rather more wide ranging. Consequently, an attempt has been made to produce a more general outline of the history and development of maps and mapping, with the efforts of civilisations being given precedence over those of individual nations.

1.2 Mesopotamia and Egypt

The earliest references to the practices of surveying and mapping are to be found on clay tablets, in papyri, and as inscriptions on the walls of tombs in Mesopotamia and Egypt. Unfortunately, apart from some tiny fragments, maps have not survived. A clay tablet of c. 3800 BC, found near Nuzi in modern Iraq, is generally recognised as being the world's oldest map. It depicts the northern part of Mesopotamia and includes the river Euphrates, the Zagros mountains in the East and the Lebanese mountains in the West. As early as 5000–3000 BC the ancient civilisations of the Middle East had perfected local cadastral systems for use in the administration of the primarily agricultural society. It might be thought that the first 'cadastre' was developed many centuries later in France, but a study of ancient history teaches otherwise! Clay tablets dating from c. 2000 BC and showing parcels of land in Mesopotamia, and papyri prepared by Egyptian land surveyors of c. 1500 BC, further demonstrate the applications of surveying and mapping techniques in a highly developed society. Engineers used plans for construction and exploration as is proved by an Egyptian papyrus of 1300 BC which illustrates a Nubian gold mine. This example, kept in Turin, is the only extant cartographic representation dating from this period.

Fig. 1.3(a) Babylonian clay tablet of c. 3800 BC, found near Nuzi in Iraq (British Museum, London)

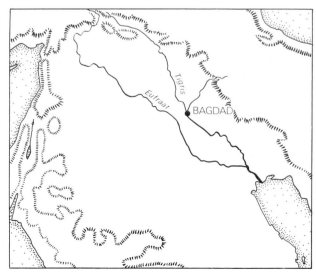

Fig. 1.3(b) Part of a modern map illustrating comparability with Fig. 1.3(a)

1.3 China and Japan

The excavation, in the People's Republic of China, of tombs constructed during the Han Dynasty has led to the discovery of three maps drawn on silk and originally produced in about 168 BC. One was a general, small-scale, topographic representation of the region; another was designed for military purposes; and the third consisted of a plan of a prefecture seat. The military map employed symbols to illustrate the hydrology, relief, communications and settlements, and is drawn at a scale of 1:90 000. Military installations are shown by means of coloured symbols, and the document should therefore properly be termed a thematic map. The regions represented are located in the south-central part of present-day Hunan Province. More recent excavation of Han tombs in Inner Mongolia has revealed several town plans, painted as murals, which can be dated to the 2nd Century BC. Evidently their compilers had considerable spatial awareness, and the mapping abilities of cartographers from the Han Dynasty certainly exceeded those demonstrated by other civilisations of the time.

Fig. 1.4 Chekiang Province from the *Atlas of China* by Chu-Ssu-pên (1273–1340). Printed from a woodcut in c. 1555 (Museum Meermanno Westreenianum, The Hague)

One of the principal advantages possessed by the early Chinese was a high level of technological knowledge and acumen. In particular, their invention of paper-making, in 105 BC, contributed enormously to the development of the arts, sciences and also administration. In 267 AD an instructional manual on surveying and mapping was prepared by P'ei Hsui, the then minister of public works. It contained instructions for the compilation of an 18-map series depicting the Chinese Empire. This was based on a rectangular grid system, and revised atlases of the country, produced under later emperors, continued to employ the same concepts. The best known of these is the *Mongol Atlas* by Chu Ssu-pên (1311–1312), which was eventually transcribed by Western scholars to produce the first atlas of China. The printing of this was undertaken in Amsterdam. The Chinese were also among the world's foremost practitioners of navigational science, with their junks visiting the coast of India and the Arabian Sea. A chart relating to navigation in Indian waters can be dated to 1416 and similar documents on rolls, up to six metres in length, are known to have been produced in the 16th Century and to have remained in use as late as the 18th Century.

In their representation of the shape of the known, habitable world, the Chinese adhered to Buddhist beliefs which are illustrated by a world map brought from India in about the 7th Century. This so-called *Map of the Five Indies* portrays a triangular shaped world surrounded by a world ocean. The concept was also introduced to Japan where it was incorporated in mapping produced as recently as the 18th Century.

Buddhist and Japanese maps based on formal, organised surveys have existed since the 8th Century. The oldest of the latter, compiled in the mid 700s, are cadastral in character and depict temple estates. This period was followed by that of the Gyōki maps which illustrated the whole of the country by means of highly distorted, round representations of each of the provinces. During the 17th Century the complex shape of Japan gradually assumed its correct geographical form. Excellent products, painted on rolls and screens, show the country in both pictorial and 'modern' styles. A cartographic milestone was reached in about 1810 with the amazing large-scale survey undertaken by Ino Tadataka. This superb project could only be improved by the application of more modern Western technology which was introduced into Japan in the second half of the 19th Century.

Fig. 1.5 Buddhist world map representing Jambu-Dvipa (the Five Indies). Drawn in Japan during the second half of the 17th Century. Size 138 x 155 cm (Courtesy of N. Muroga)

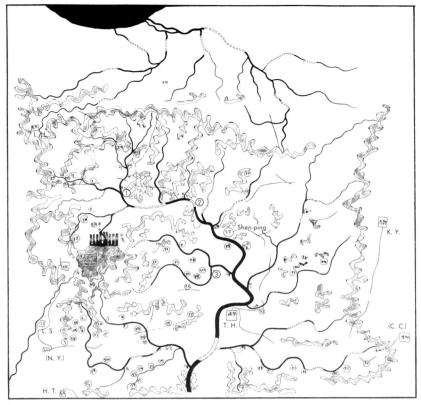

Fig. 1.6(a) A topographic map drawn on silk and dating from c. 168 BC.
Excavated from a Han Dynasty Chinese tomb

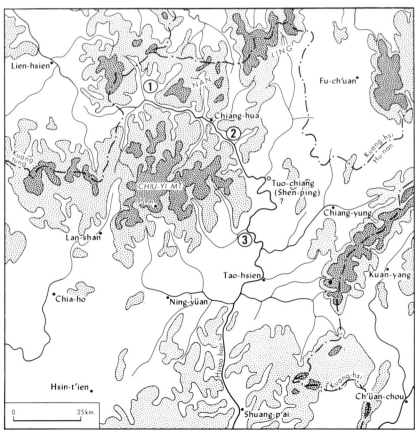

Fig. 1.6(b) Part of a modern map illustrating similarities with Fig. 1.6(a)

10

1.4 Cartographic knowledge of the ancient Greeks and Romans

Research has proved that, since c. 400 BC, the Greek intellectual classes had been capable of mapmaking. Writings from this period suggest that cartography was actively practised by Greek philosophers, astronomers and mathematicians, but that the role of the land surveyor was purely incidental. People of this time had a reasonable appreciation of the characteristics of the remainder of the world, as is demonstrated (in terms of both its shape and size) by a consideration of maps generated by Eratosthenes (278−175 BC). The content of these is based on information obtained from independent travellers. An even more extensive knowledge is exemplified by the world map which incorporates details collected by Claudius Ptolemaeus (Ptolemy) of Alexandria (87−160 AD). Here, for the first time in the history of mapmaking, a geographical graticule system comprised of meridians of longitude and parallels of latitude was employed. All of the places in the world of which Ptolemy was aware were listed in his book entitled the *Geographia*, which was compiled in about 150 AD. This was later copied by Arab writers during the 9th Century, and may have influenced Moslem geographers, although no copies in Arabic have survived.

The *Geographia* was also copied by Byzantine scribes (the oldest surviving manuscript dates from the 12th Century), and the work has had an extremely important and long-lasting effect on Western cartography. However, it is debatable as to whether Ptolemy ever actually drew the twenty-seven maps with which he is credited—all of which are incorporated in 12th−15th-Century manuscript editions—and it is thought that they may have been compiled from the original, highly descriptive text during the 12th Century.

Unlike the Greeks, the Romans were not concerned with the scientific basis of mapping, but prepared documents for more practical reasons (administration, military purposes, and to aid more efficient communication within their extensive Empire). Unfortunately, only fragments of large-scale, mosaic, cadastral and town plans (dating from the 1st Century) still exist. Manuscript versions of Roman writings and maps, originally produced on parchment, have only been preserved in a copied form since Mediaeval times. The best known of these is the *Peutinger Table* which is named after its 16th-Century owner. It is a road map relating to the whole Roman Empire and, although now incomplete, was produced in a rolled form measuring 6.75 x 0.34 metres overall. Maps designed to fit a long, narrow strip of parchment must have been very common during the time of the Roman Empire. Although they were produced in a readily portable form, the representation of either the known world or a specific region within it was obviously extremely distorted.

1.5 Moslem cartography

Western technology owes its awareness of the sciences of surveying, astronomy and navigation to Islamic scholars who continued to work and develop expertise during the European 'Dark Ages'. This knowledge gradually spread, via Arab universities in Spain, to France, Italy, Austria and Germany. Later, during the European Renaissance, Moslem science and technology became combined with Greek mathematical and Roman practical talents. After about 1600 the academic pendulum of the history of civilisation swung back, and Western traders initiated a period of conquest and colonisation in Asia which had previously served as the 'cradle' for the development of European science and technology.

Because of the vast geographical extent of the Arab Empire, good mobility was essential, and thus there was a need for accurate and reliable maps showing communications. Roadbooks, itineraries and geographical maps of outstanding quality were compiled by various authors from about 850 AD. These were collected together and assembled, in book form, as the *Atlas of Islam*. The work was comprised of twenty-one maps illustrating the Mediterranean, Arabian Gulf, Syria, Iraq and the Caspian Sea and maintained a very consistent form and content for some three generations. After about 1150, when the Norman peoples had spread throughout Europe, formal cultural, commercial and political relations were established with the Moslem world. The Arab geographer Idrisi practised cartography at the court of King Roger II on the isle of Sicily, and his merging of information recorded in the *Atlas of Islam* with Norman knowledge resulted in a general improvement of geographical awareness. In about 1150 Idrisi produced both a world map and an atlas, and subsequently, in about 1160, he wrote a geographical text which was illustrated with maps. Copies have survived.

The Arabs, Turks and Norman peoples were all excellent sailors and ruled the seas. The earliest nautical charts were produced in countries bordering the Mediterranean in about 1250, and it is quite possible that the characteristic shape and style of these followed conventions originally established in Idrisi's time.

1.6 Commercial motives for mapmaking

As has been the case with all crafts, map drawing was also practised for commercial reasons and this is a very evident characteristic of European

Fig. 1.7(a) World map by Ibn al Virdi (1292–1349). See:
Conrad Miller's *Mappae Arabicae* Vol. V, 11, London 1,
fig. 78.11 (Stuttgart, 1931)

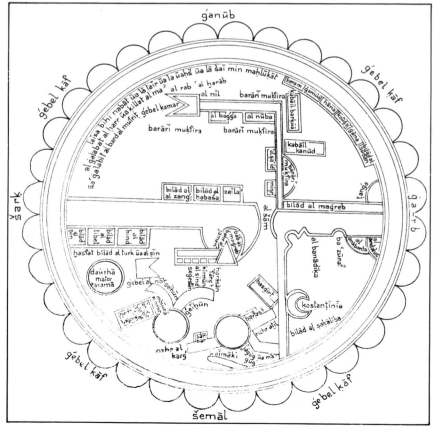

Fig. 1.7(b) Redrawn and transcribed version of the above map

12

cartography. It became particularly notable following the effecting of three major technological developments: the introduction of paper-making to the continent; the application of woodcuts and copper-engraving in the production of printed illustrations; and finally, in about 1450, the invention of book printing techniques. Contrary to the role of mapmaking in the Moslem world between the 9th and 15th Centuries, and in the Sino-Japanese and Indian civilisations before the 19th Century, Western cartography served to satisfy commercial rather than political motives. The professional European chartmaker, undertaking commissions for clients, can be traced back to the 14th Century. Evidence exists from c. 1300 of a contract for the preparation of world maps of considerable dimensions by cartographers working in Barcelona.

As a natural consequence of the European nations' pursuit of expansionist policies from the 13th Century onwards, a great deal of information relating to the remainder of the world became available. This knowledge was immediately commercially exploited by scribes, painters and later printers. During Europe's 'Great Age of Exploration and Discovery' (c. 1450–1650), map generation was greatly assisted by the availability of an ever-increasing supply of new geographical information. Various printers and publishers, often from different countries, vied with one another in the fiercely competitive production of world atlases, a number of which went into several editions. A major selling point was the incorporation of newly mapped details relating to Asia, America and Africa. The Dutch proved to be pre-eminent in the development of commercial atlas preparation during the 17th Century, and published nautical atlases, pilot books and world atlases. The first regularly produced world atlas was issued in Antwerp (then part of the Southern Netherlands) in 1570, and the largest work of this type ever published, the *Grand Atlas*, emanated from Amsterdam in 1664.

Turkish and Arab world atlases can be dated back to the 10th Century, but appeared only in manuscript form. They were not produced for commercial reasons but were compiled, by order or command, to illustrate the world as it was known to the Moslem school of geographers. These maps, although much better than those of the same area produced by European cartographers prior to 1700, had no influence on early Western mapping.

The history of Chinese cartography is extremely interesting, as is the influence that it exerted on the appearance of Western-generated maps of the country. After mapmakers had gained some thousand years of experience in the generation of maps of China, the so-called *Mongol Atlas* (mentioned earlier) was compiled by Chu Ssu-pên

Fig. 1.8 A state book-case holding the 11-volume *Grand Atlas* produced by Joan Blaeu in Amsterdam, 1664 (University Library, Amsterdam)

(1273–1340). It appeared in several editions, with the original appearing as a manuscript version and those produced after 1550 exhibiting a printed form. This work remained the definitive atlas of China until it was revised by Jesuit (Christian) missionaries in the 18th Century. A Latin edition, with the maps redrawn in the European style, was prepared by the Jesuit Father Martin Martini and printed in Amsterdam by Joan Bleau in 1655. This *Atlas Sinensis* was the first atlas of China to be commercially produced in the Western world. Although the Chinese had produced their 'national atlas' some six hundred years before any of the European countries attempted such a project, the Jesuits were able to demonstrate that the accuracy of the maps included could be improved. They did this by the implementation and direction of more precise surveys, and the employment of Western instrumentation. These surveys provided the basis for a French edition of the *Atlas de la Chine*, which was compiled by J. B. d'Anville and printed in Paris in 1730.

1.7 Style and composition

Generally speaking, it is true to say that during its infancy cartography demonstrated a rather more pictorial style than was the case in the 19th Century or today. The time of transition occurred, in the Western world, in the late 18th Century, which witnessed the heyday of the military topographers (ingénieurs géographes militaires).

Fig. 1.9 Part of a map of Russia by D. Gerasimov (1525); from a manuscript atlas by Battista Agnese (Bibl. Marciana, Venice)

Between c. 800 and 1800 a pictorial approach was adopted for the representation of the terrain, settlements, vegetation, the oceans and their coasts, as is shown by Western maps and also those produced by the Japanese and Koreans during the same period. Conversely, pictorial elements are conspicuously absent from works generated by Islamic cartographers at a similar time. The premier reason for this eventual switch from illustrative to abstract symbolisation results from the increasing need for the use of maps in technical situations — the construction of fortifications, building of towns, etc. — as well as for administrative purposes (cadastral plans).

Prior to 1800, buildings were conventionally portrayed 'in elevation', even on the ground plans relating to property surveys, and town plans (normally incorporating elements of the surrounding countryside) were produced as 'bird's-eye' views. Typically nautical charts contained very graphic coastal profiles and inland detail, together with decorative items such as sailing ships, sea monsters and mermaids. Commercially available maps, printed from copper-plates, were typically embellished with elaborate, decorative lettering, human figures, and detailed borders which included heraldic devices, small town views, etc. A map, whether it be of Western, Asiatic or Islamic origin, can be approximately dated by a consideration of its style and content. However, it should be noted that only experts are sufficiently skilled to establish the true age and origin of early cartographic documents.

1.8 History of cartographic techniques

The development of map production technology is of considerable interest to modern cartographers. Although manuscript maps and charts were originally manually generated, advances in civilisation eventually resulted in a necessity for the publication of multi-copies of individual documents for a variety of practical and commercial reasons. For example, thousands of copies of contemporary navigational charts were required when the huge 17th-Century European sailing fleets 'swarmed' into the Atlantic and Indian Oceans. Originally, linework, lettering and colouring were produced freehand, as were facsimile copies of manuscript documents. The first map-printing press was installed in Turkey in the 18th Century; and China, Japan and Korea used wood-blocks to reproduce illustrations in the early 16th Century. The first West European map was produced from an engraved copper-plate in 1477, but it was not until the 17th Century that the Jesuits introduced this printing technique into China. Commercial map production using copper-plates was quickly adopted in Europe, but the results were monochrome and any additional colouring had to be added by hand. This resulted from the difficulties

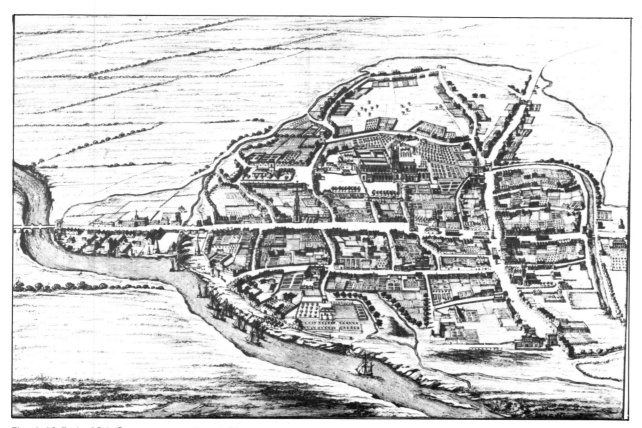

Fig. 1.10 Early 18th-Century town plan of Gloucester; engraved in 1710 by John Kip (Bodleian Library, Oxford)

15

Fig. 1.11 Part of the original wood block used in printing a sheet of Philipp Apian's *Bayerische Landtafeln*, 1568 (Bayerisches Nationalmuseum, Munich)

Fig. 1.12 Copperplate engraving and printing. Engraved by Philip Galle from a drawing by Jan van der Stratt (1560). A = senior engraver (note glasses); B = apprentice; C = plate preparation; D = press operation; E = print drying (Plantijn Museum, Antwerp)

inherent in registering and printing area tinting with previously produced linework, which necessitated the use of a further plate for the application of each additional colour required. Children were employed as map colourists in the 17th- and 18th-Century European publishing houses. An 18th-Century encyclopaedia informs us that, 'In Augsburg children used to be employed for the illumination of maps. They were required to complete a hundred copies in three days, and for this were paid 18 Groschen.' It is possible to produce multi-colour maps from wood-blocks, as is demonstrated by a number of beautiful maps printed in Japan during the 18th and 19th Centuries.

ment of the fast-running offset litho press. After 1940 the availability of plastics for use as transparent bases for map drafting led, initially, to some decrease in graphic quality. However, from 1960 the standard of line-work again improved as a result of the introduction of new engraving techniques for use with dimensionally stable, coated, polyester-plastic sheets. Scribe-coatings, scribing tools, strip-masking, phototypesetting, etc. — all inventions of the last three or four decades — have given the map an appearance of graphic perfection. Nevertheless, it should be noted that the manual skills required to achieve this degree of excellence are not nearly half as great as they were in the 19th Century!

Fig. 1.13 Copper engravers working a 19th-Century studio

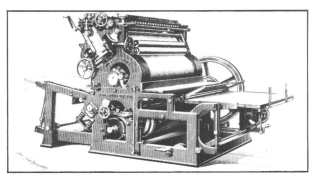

Fig. 1.14 An early 'Algraphia' rotary press built, prior to the introduction of offset, by Bohn and Herber, Würzburg, Germany in c. 1880

At the end of the 18th Century, a Bavarian, named Alois Senefelder, developed the lithographic method of printing which involved the transfer of an image from stone onto paper. By using this process, coloured areas and symbols could be easily reproduced and, in consequence, lithography quickly became the most important map production process. From about 1840 multi-colour printing became common for the generation of large quantities of documents, but hand-tinting was still used for short runs. The process is still in common use, but has now been developed into what is called 'offset lithography'.

A second, vitally important invention relating to the development of cartographic technology was that of photography. Light-sensitive chemicals were employed in the transferring of a line-drawing onto a printing plate, or were used in the making of duplicate copies of an original map at the intended publication scale. The first process camera became available in about 1800, and subsequently it became unnecessary to draw an original map at publication scale. In consequence, cartographers gradually lost their talents as miniaturists!

Further advances occurred in about 1900 with the introduction of zinc (and later aluminium) plates as replacements for the heavy and unwieldly lithographic stones, and also with the develop-

1.9 Map types

In subsection 4.2.2 of Chapter 4 of Volume 2 of *Basic Cartography* the authors of information on Thematic Cartography list a wide selection of different types of maps based on variations in their subject matter. Did all of these map types have precursors in earlier times? The answer to this question is no, because the necessary scientific developments relating to the extension of human knowledge with respect to detail appropriate for use in mapping are comparatively recent.

The oldest regional geological maps date only from c. 1810; the classification of soils in areas of limited size from c. 1840; and information on climate, hydrology, geomorphology, vegetation, etc., over large areas was first collected, between 1800 and 1820, by Alexander von Humboldt, the founder of modern physical geography. A search for earlier examples of maps relating to aspects of our physical environment is pointless!

What about socio-economic and demographic mapping? As soon as the human race began to multiply to the extent that its numbers reached significant proportions, scientists became interested. Methods for the collection and analysis of statistics were introduced in the early 1820s and, based on somewhat spurious detail, the first

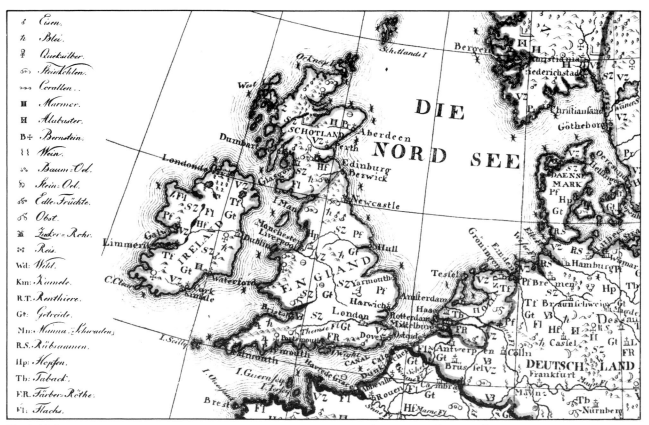

Fig. 1.15 Section from the earliest known, printed, thematic map — A. F. W. Crome's *Neu Carte von Europa welche die Merkwürdigsten, Producte und Vornehmsten Handelsplätz . . . enthält.* Engraved by F. A. Pingeling, Hamburg, Dessau, 1782 (Geografisch Instituut, Utrecht)

socio- and economic-statistical maps were produced in about 1830. At that time the world population was of the order of one-sixth of its current size! Since 1940, industry, trade and commerce, and man's increasingly complex methods for the use of minerals and natural resources, have led to significant increases in the numbers and varieties of published maps. In conclusion, the planning and control of man's future use of limited living space, food supplies, energy and water resources will, inevitably, result in the necessary creation of new map types which will, it is to be hoped, be of historical interest to cartographers of the year 2500.

Chapter 2

MATHEMATICAL CARTOGRAPHY

D. H. Maling

CONTENTS

THEORY OF MAP PROJECTIONS

USEFUL MODERN REFERENCES ON MATHEMATICAL CARTOGRAPHY

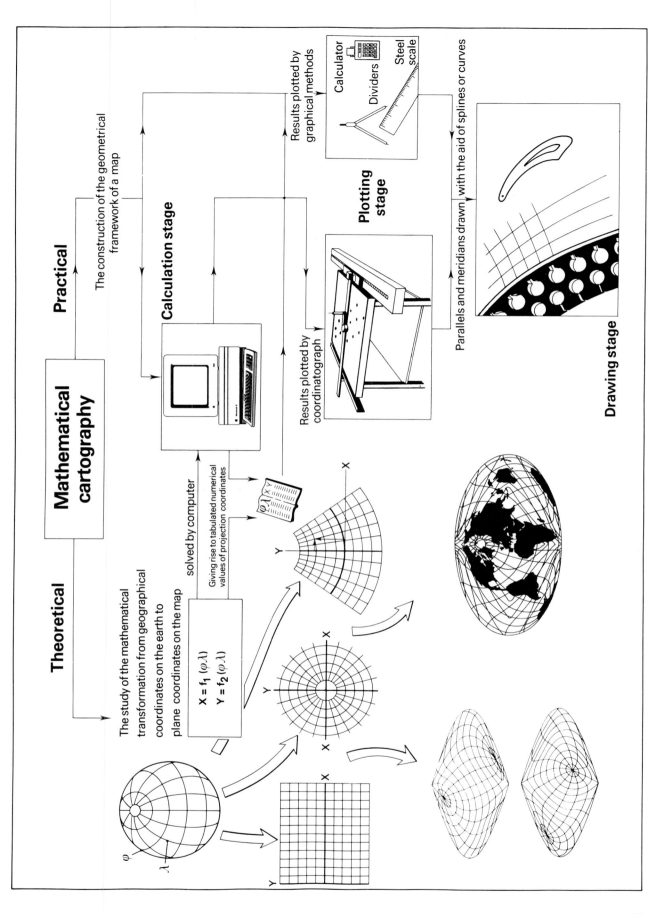

Theoretical

Practical

Mathematical cartography

The study of the mathematical transformation from geographical coordinates on the earth to plane coordinates on the map

The construction of the geometrical framework of a map

$X = f_1 (\varphi, \lambda)$
$Y = f_2 (\varphi, \lambda)$

solved by computer

Giving rise to tabulated numerical values of projection coordinates

φ, λ, x, y

Calculation stage

Results plotted by coordinatograph

Results plotted by graphical methods

Calculator

Dividers

Steel scale

Plotting stage

Parallels and meridians drawn with the aid of splines or curves

Drawing stage

Fig. 2.1

21

DEFINITIONS

2.1 Introduction

Mathematical cartography is that branch of cartography concerned with the mathematical basis of map making, particularly the study of map projections.*

The practical aspects of the subject include the measurements, calculations and the plotting procedures which are used to plot and draw the map projection which has been chosen to serve as the geometrical framework for a particular map, as a network of lines upon the map manuscript. This construction is the essential preliminary to the compilation and fair drawing of the detail to be shown on the map, because all of this detail must be fitted within the geometrical framework in order to produce a map of required scale and possessing definite mathematical properties.

The theory of mathematical cartography is mainly concerned with the investigation of the ways in which the curved surface of the Earth may be portrayed, by means of a map projection, on the plane surface of a map. An infinitely large number of possible ways of doing this may be described. The practising cartographer should know something of the mathematical properties of map projections and how to make a reasonable choice of projection for a particular map.

2.2 Terms used to describe the mathematical framework of a map

We study the use and meaning of the following words which describe the geometrical framework of a map: Scale ①; graticule ②; grid ③; neat lines ④; and sheet numbering system ⑤.

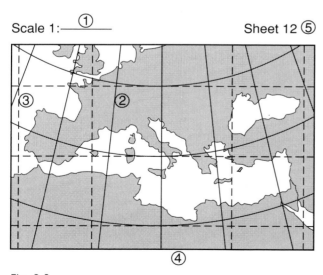

Fig. 2.2

2.2.1 Scale

The scale of a map is the ratio of distance measured upon it to the actual distances which they represent on the ground.

For example, if a straight line on the ground of length 2.5 km is shown on a map by a line of length 2.5 cm, the scale of the map may be calculated from:

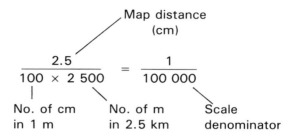

Once we have determined the scale of a map, we may use this information in two different ways:

1. If the scale of the map is known to be 1/25 000, what is the map distance which corresponds to a ground distance of 2 km?

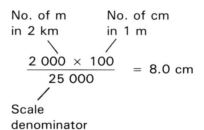

This is the scale conversion most commonly used by the cartographer.

2. If the scale of the map is known to be 1/50 000, what is the ground distance which corresponds to a map distance of 3.0 cm?

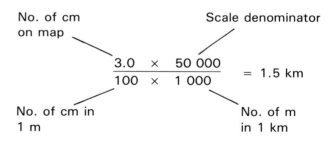

This is the scale conversion most commonly made by the map user.

The two examples illustrate the scale conversions used to determine map measurements from ground measurements and the reverse. In making

---

*Definitions in bold type given in this chapter are those used in the ICA *ICA Multilingual Dictionary of Cartographic terms*, Wiesbaden, 1973.

these calculations, **all** the measurements must be expressed in the same units. In each case the ground distances must be converted into centimetres to correspond with the map distances.

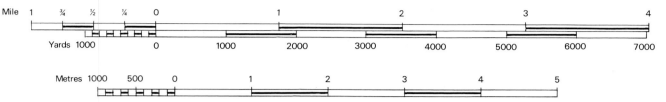

Fig. 2.3

2.2.1.1 The graphic scale
The scale of a map is usually shown by means of one or more graduated straight lines which are subdivided into units of ground distance.

2.2.1.2 The Representative Fraction
This is **the scale of a map or chart expressed as a fraction or ratio which relates unit distance on the map to distance, measured in the same units, on the ground.**

The numerical value for scale may be written in various ways, all of which have the same meaning:

       1:25 000   1:25,000
       1/25 000   1/25,000

$$\frac{1}{25\ 000} \qquad \frac{1}{25,000}$$

The use of the comma (,) in the right-hand column is usual British and American practice and it should not be confused with the symbol for a decimal.

2.2.1.3 Classification of maps by scale
Three terms are frequently used to classify maps according to scale. The actual scale values attributed to each class depends upon the customary usage by particular publishers or government departments. That suggested by ICA is: large-scale maps—scale larger than 1/25 000; medium-scale maps—1/50 000—1/100 000; small-scale maps—scale less than 1/200 000.

2.2.2 Graticule

The graticule is a network of lines shown on the body of the map and sometimes by subdivision of the neat lines or border of the map. One family of these lines represents the **parallels of latitude** (2.4.1); the other family represents the **meridians of longitude** (2.4.3).

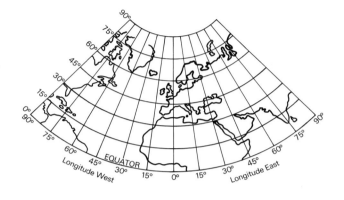

Fig. 2.4 Equidistant conical projection with one standard parallel (30°N)

Each graticule is based upon a particular map projection, and according to the choice of projection:
- —the lines may be straight or curved;
- —the lines may be parallel or converging;
- —the separation of the lines may be constant or vary from place to place;
- —the angle formed by the intersection of a parallel and a meridian may be of any size.

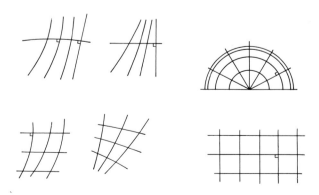

Fig. 2.5

2.2.3 Grid

A grid on a map is a system of straight lines intersecting one another at right angles. It represents a method of defining position on the ground by means of distances measured upon a plane surface which is assumed to correspond to a portion of the Earth's surface. Many countries are mapped upon one or more local grids which have been devised especially for that purpose. In addition, there are some grid systems which react with most of the Earth's surface in a systematic fashion, e.g. the Universal Transverse Mercator (UTM) Grid. See also 2.5.4.

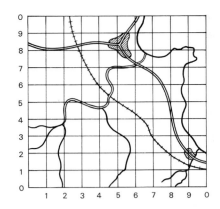

Fig. 2.6

2.2.4 Neat lines

The neat lines of a map are **those which enclose all the map detail and therefore define the limits of the area mapped.**

Three kinds may be encountered:

— On large-scale and some medium-scale maps the neat lines are grid lines. Consequently, the format of the map is always square or rectangular.

— On small-scale maps and a few medium-scale maps the neat lines are formed by two parallels and two meridians of the graticule. These may be straight or curved lines and frequently the edge of the map closer to the geographical pole is shorter than the side nearer the Equator.

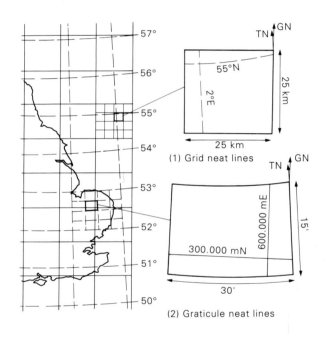

Fig. 2.7 (1) Grid neat lines, and (2) Graticule neat lines

— The neat lines are arbitrary straight lines which have no relationship to either grid or graticule, and merely serve to subdivide the area to be mapped into a series of rectangular maps of similar dimensions. Arbitrary neat lines seldom fit an irregular shaped country in a convenient way. The illustration shows some of the methods which are used to try to fit a country into the minimum number of map sheets. These problems also arise in atlas cartography because most atlas maps have arbitrary neat lines.

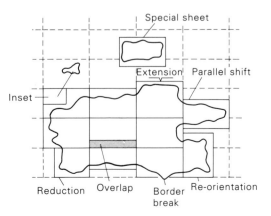

Fig. 2.8 Arbitrary neat lines

2.2.5 Representation of the grid, graticule and neat lines on maps at different scales

Table 2.1

Scale	Grid separation (km)	Graticule separation (° or ')	Neat lines
1/5 000 or larger	0.1	—	Grid
1/10 000	1.0	—	Grid
1/25 000	1.0	—	Grid*
1/50 000	1.0	5'	Grid*
1/100 000	1.0	5'	Grid*
1/250 000	10.0	5'	Grid or graticule
1/500 000	10.0	30' or 1°	Graticule
1/1 000 000	—	1°	Graticule
1/5 000 000	—	5°	Graticule or arbitrary
1/20 000 000	—	10°–5°	Graticule or arbitrary

*Certain important exceptions occur. For example, the topographical maps published by the US Geological Survey at 1/24 000, 1/62 500 and 1/125 000 have graticule neat lines.

2.2.5.1 The appearance of the grid and graticule lines

This varies according to the mapping specification. For example:
- The colour of the grid may be the same colour as that of the Primary Plate, or it may be different from any of the other colours used on the map.
- On maps which show the overlap of two different grid systems, these are normally distinguished by printing them in different colours.
- It is common for every tenth grid line to be drawn with a wider line than that used for the others.
- Although grid and graticule lines are normally shown on modern maps by continuous lines, sometimes a **rouletted** grid or graticule may be seen. This is a method of depicting the lines by means of small, equidistantly-spaced dots.
- On maps of scale 1/500 000 and larger, the graticule may only be shown by small crosses and by subdivision of the neat lines. The crosses on the map may be omitted where they coincide with other map detail.
- The graticule is usually drawn on the Primary Plate and all graticule lines are shown with equal width.

2.2.6 Sheet numbering systems

The majority of maps form part of a series. Each map is one from the hundreds, or even thousands, of topographical maps needed to cover a whole country. It may be an individual page from an atlas. In order to aid identification, each map sheet or page is numbered and it may also have a name. Generally, the name given to a map sheet is descriptive of the area as a whole (such as the name of a country or continent on an atlas map) or it is the name of the most important feature (town, mountain, lake) on that topographical map. Three different systems of sheet numbering may be used, depending upon the nature of the sheet lines used for the whole series.

2.2.6.1 Grid reference designation

If the neat lines are grid lines, the sheet numbering system is usually based upon the method of making a grid reference. Usually it is the grid reference for one corner of the map and frequently the south west corner is the point to which the numbers refer.

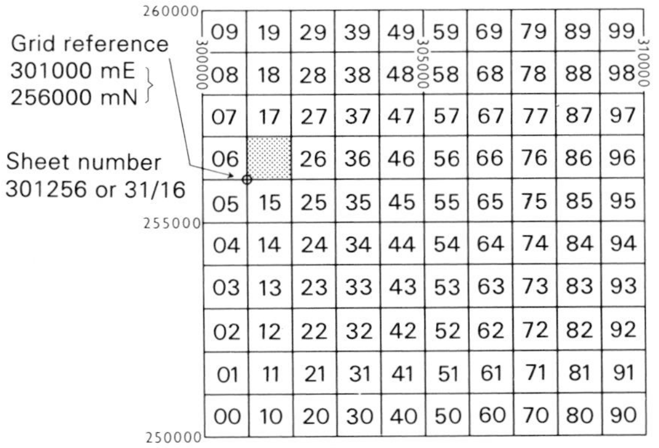

Fig. 2.9

2.2.6.2 Arbitrary numbering systems

If the neat lines are arbitrary boundaries, the maps are numbered serially. This is frequently done along west–east zones beginning in the north west and ending in the south east. In some countries, however, the zones run north–south. In each case, sheet 1 is located in the north west corner of the area mapped and the map with the largest serial number is in the south east.

1	2	3
4	5	6
7	8	9
10	11	12
13	14	15
16	17	18
19	20	21
22	23	24

1	9	17
2	10	18
3	11	19
4	12	20
5	13	21
6	14	22
7	15	23
8	16	24

Fig. 2.10

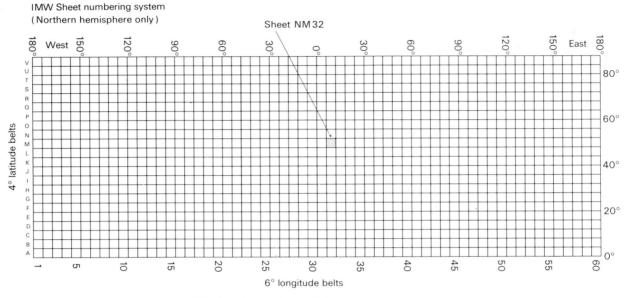

Fig. 2.11 IMW sheet numbering system (Northern hemisphere only)

2.2.6.3 The sheet numbering systems of IMW

If the neat lines are elements of the graticule, the sheet numbering system may be based upon the geographical co-ordinates of the south west corner of the map. However, the most common numbering system, which uses latitude and longitude, is based upon that adopted for the International Map of the World (IMW) at 1/1 000 000 scale.

Sheet designation is made by using the following letters and numbers:
- The letter N or S indicates whether the sheet lies in the Northern or Southern hemisphere.
- The latitude zone of the map is indicated by a letter A through to V. Since each map extends 4° in latitude, each of these zones measures 4°. Thus A represents the zone 0°−4°, M the zone 48°−52° and V the zone 84°−88°.
- The numbers 0−60 are used to denote the longitude zones, each of which measures 6° and corresponds to the west−east extent of the single map. The numbering of the zones is eastwards from 180° longitude, so that 0 represents the zone 180°−174° West (of Greenwich), 32 the zone 6°−12° East and 60 the zone 174°−180°.

Sheet identification is composed of these three elements in the order described. For example, the 1/1 000 000 map bounded by the parallels 48°−52° North and by the meridians 6°−12° East has the number, N M 32.

Many countries have adapted this system for further subdivision to create sheet numbering systems for map series within the scale range 1/50 000−1/500 000.

2.3 The shape and size of the Earth

Knowledge about the Earth's shape and size (known as the **Figure of the Earth**) is essential if we are to make maps of its surface. It is necessary to know about its size in order to make maps of it at known scale. The shape of the Earth's surface influences the kind of mathematical transformation required to map it onto a plane surface.

We study the meaning of some of the following key words: geoid; sphere; ellipsoid of rotation; spheroid; flattening; radii of curvature.

2.3.1 The geoid

Detailed knowledge about the Earth's figure has

been obtained from a variety of sources: geodetic surveys; studies of variations in gravity; astronomical methods, especially by tracking the orbits of artificial satellites.

All these methods define a slightly irregular surface known as the **Geoid**. The principal difference between the geoid and a perfect sphere is a flattening of the geoid towards the geographical poles. Other gentle undulations are important to the study of geodesy and geophysics, but may be neglected in cartography.

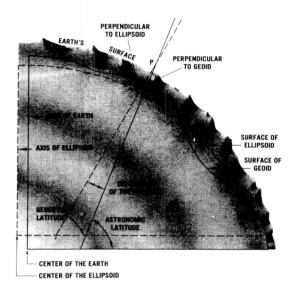

Fig. 2.12

Because of the polar flattening, the geoid corresponds fairly closely to an **Ellipsoid of Rotation** with an equatorial radius (major axis) of about 6 378 km and a polar radius (minor axis) of about 6 357 km. This figure may be illustrated in section by an ellipse.

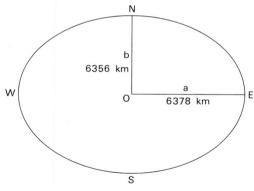

Fig. 2.13

The flattening of an ellipsoid may be defined by: $f = (a - b)/a$, and is expressed as the fraction $1/f$.

For the Earth, $1/f = 1/298$. An ellipsoid with such small flattening may also be called a **Spheroid**. Since the polar axis is shorter than the equatorial axis, we may further describe it as an **Oblate**

Spheroid. The diagram shows that even flattening of 1/50 gives rise to an ellipse which is nearly circular. Thus all the elliptical diagrams which illustrate the spheroid are much exaggerated.

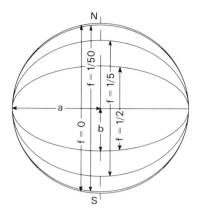

Fig. 2.14 Scale drawing of a circle with radius a, and ellipses with major semiaxis a, having different amounts of flattening. The circle has flattening $f = 0$. Even an ellipse with $f = 1/50$ is nearly coincident with the circle. The diagram illustrates that an ellipse with $f = 1/298$, corresponding to the Earth's reference figure, cannot be distinguished at this small scale from a circle.

2.3.2 The geometry of the sphere and spheroid

We may compare the geometry of the two different bodies as follows.

The Sphere

1. All points on the surface of a sphere are equidistant from its centre. Therefore, any straight line which joins the centre, O, to any point, P, on the surface represents the radius, R.

2. Any plane section passing through the centre of a sphere may be represented by means of a circle with radius, R. This is known as a **Great Circle**. Any plane section passing through the sphere, but not through its centre, may be represented by means of a **Small Circle** having a centre, O', and radius, r, which is less than R.

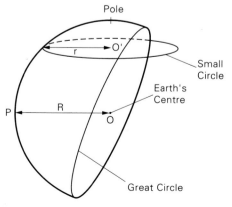

Fig. 2.15

3. The arc distance, AB, on the spherical surface is measured by the angle, AOB = z, formed at the centre of the sphere between the two radii drawn to the points A and B. If z is measured in radians, AB = Rz.

It follows that for any given value of z, the length of the arc, AB, is constant, irrespective of its position on the surface of the sphere.

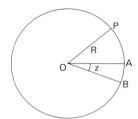

Fig. 2.16

4. A sphere has only one radius, R.
5. Any tangent to the surface of a sphere, such as the line, AT, is perpendicular to the radius drawn to the point of contact, i.e. angle TAO = 90°

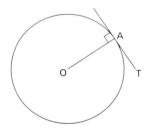

Fig. 2.17

The Spheroid
1. Points on the surface of a spheroid lie at different distances from its centre. The extremes are OE = a, which is the greatest, and ON = b, which is the smallest.
2. A plane section through the centre of a spheroid is, with one exception, an ellipse. The exception is the equatorial section, through the point E and W, which is a circle with radius a.

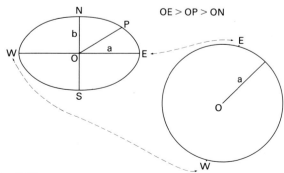

Fig. 2.18

3. The length of the arc corresponding to the angular distance, z, varies in different parts of the spheroid. Thus the arc, AB = z, which is near the Equator, is shorter than the arc, CD = z, which is near the geographical pole.

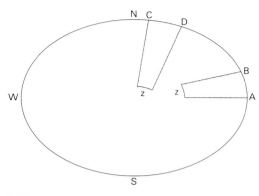

Fig. 2.19

4. A spheroid has two radii of curvature at each point on the surface, and these radii vary continuously from place to place.
   —**Meridional Radius of Curvature**, ρ, which is the radius of the elliptical section along the meridian NAE through the point A. Note that the line corresponding to this radius is AQ', which does not pass through the centre of the spheroid.
   —**Transverse Radius of Curvature**, ν, which is the radius of the elliptical section which also passes through A, but is perpendicular to the meridian. This radius corresponds to the line AQ.
5. The line which is perpendicular, or **normal**, to the tangent AT does not pass through the centre of the body. In the diagram it is the line AQ and not the line AO.

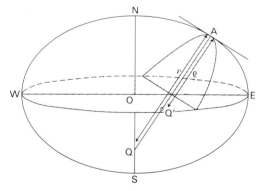

Fig. 2.20

2.3.3 The spherical assumption and the spheroidal assumption

The sphere and the spheroid are geometrical bodies which only approximate to the shape of the geoid. The spheroid fits the geoid better than does the sphere, but it is mathematically more complicated and the use of it as a **reference figure** involves some long-winded and quite difficult calculations. The geometry of the sphere is much simpler, but the use of this assumption is less accurate.

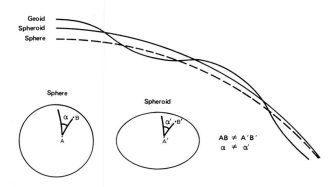

Fig. 2.21

If we consider two corresponding points on the surface of the two bodies we can compute the lengths of the arcs AB and A′ B′, and also the direction and angle which these arcs make with a fixed direction. There is a small difference between the two lengths and the two directions. If these discrepancies can be detected on a map, i.e. if B and B′ are more than 0.2 mm apart (assuming A and A′ to be fixed in position), it will be desirable to use the spheroidal rather than the spherical assumption about the shape of the Earth.

This normally applies to large scale topographical mapping. The need to use the spheroidal assumption at the larger scales is further reinforced by the fact that these are maps based upon original surveys. These are adjusted and computed in natural ground units (i.e. at the map scale 1/1). For this part of the work it is essential to determine points on the spheroid which have been adopted for use in a particular country.

At small map scales the discrepancies between the sphere and spheroid are so small that they may be neglected. It is therefore sufficient to use the assumption that the Earth is a perfect sphere for atlas and thematic maps.

2.3.4 The Figure of the Earth

No single spheroid is considered to be suitable for all surveys and topographical mapping throughout the world.

For historical and political reasons, as much as for scientific merit, about 15 different figures of the Earth are in current use.

In making the spherical assumption, the logical choice is the use of the radius of a sphere, which has either the same surface area or the same volume as a particular spheroid.

The results are similar by either method of calculation and for all the spheroids in common use. It is sufficient, for practical purposes, to regard the spherical radius of the Earth as 6 371.1 km.

2.4 Geographical co-ordinates

The best known method of referring to position on the Earth's surface is by means of the two angles, **Latitude** and **Longitude**, which together form the system of Geographical Co-ordinates. Important differences occur in the definitions of latitude as it is measured on the sphere or spheroid. The definition of longitude is the same for both kinds of reference figure.

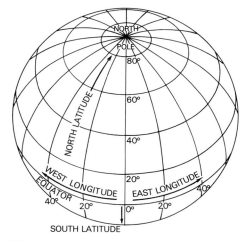

Fig. 2.22

Table 2.2

Name of spheroid	Date	Major axis (m)	Minor axis (m)	Flattening
Everest	1830	6 377 276	6 356 075	1/300.80
Bessel	1841	6 377 397	6 356 079	1/299.15
Clarke	1866	6 378 397	6 356 584	1/294.98
Clarke	1880	6 378 249	6 356 515	1/293.47
International	1924	6 378 388	6 356 912	1/297.00
Krasovsky	1940	6 378 245	6 356 783	1/298.30
International Astronomical Union or GRS67	1968	6 378 160	6 356 774	1/298.25

The details concerning national usage of these and other figures of the Earth are to be found in *World Cartography*, **x**, 1970; **xiv**, 1976; **xvii**, 1983.

We study the meaning of the following key words: latitude; geocentric latitude; geodetic latitude; co-latitude; longitude; parallels of latitude; meridians.

2.4.1 Latitude

Sphere latitude is the angle measured at the centre of the Earth between the plane of the Equator and the radius drawn to a point on the surface. In the diagram the latitude of the point, P, is the angle POE.

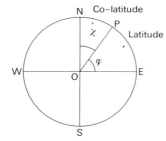

Fig. 2.23

The Equator is the datum for measurement of latitude and this is therefore assigned the value 0°. Northwards and southwards from this datum, the latitude increases until it is 90° North at the North Geographical Pole and 90° South at the South Pole. In calculations which are made using geographical co-ordinates, North latitude is reckoned + and South latitude −. Algebraically, latitude on the sphere is denoted by the Greek letter φ (phi) as is geodetic latitude on the spheroid.

Spheroid
Two different angles may be used to measure latitude:
—**Geocentric Latitude**, ψ, is the angle POE, measured at the centre of the body between the plane of the Equator and the straight line OP.
—**Geodetic Latitude**, φ, is the angle PME, measured at M in the plane of the Equator, where the normal to the spheroidal surface at P meets it.

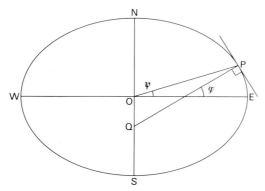

Fig. 2.24

There is a small difference between the two angles ψ and φ, and this varies with the position of P on the surface of the spheroid. Although Geocentric Latitude appears to correspond more closely with the definition of latitude on the sphere in geodesy and cartography, it is the Geodetic Latitude which is invariably used.

2.4.2 Co-latitude

In some calculations it is more convenient to use the angle NOP than the latitude POE. Since NOE is a right angle, NOP = 90° − φ, and is known as the co-latitude. This is denoted algebraically by the Greek letter χ (chi). On the spheroid, χ = 90° − φ, where φ is the Geodetic Latitude.

2.4.3 Longitude

The diagram shows two planes, both of which pass through the centre of the Earth and both of which are perpendicular to the Equator. Consequently, the two planes intersect along the axis NOS (the axis of rotation of the Earth) and their circumferences are great circles. One plane, NPS, contains the point P on its circumference. The other plane, NGS, contains a datum point, G, from which longitude is measured. Longitude may be defined as the angle, measured at the centre of the Earth, between the plane containing the point P and the datum plane. Hence, it is the angle COD. This angle is measured eastwards and westwards from the datum plane and is recorded as East longitude or West longitude. In calculations, East longitude is + and West longitude is −. The angle is denoted algebraically by the Greek letter λ (lambda). We use the letters $\delta\lambda$ to indicate the difference in longitude between two places.

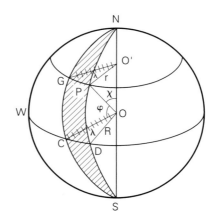

Fig. 2.25

The datum from which longitude is measured may be chosen as the plane passing through the origin of a national survey. This is how longitude is still recorded on many national topographical maps. For other purposes, especially in navi-

30

gation, it is better to use a single, internationally recognised, datum. In 1884 this datum was selected as a point at the site of the Royal Observatory at Greenwich, near London, in England. This is known as the **Prime Meridian** or **Greenwich Meridian**.

2.4.4 Parallels and meridians

The locus of all points having the same latitude traces a circle upon the spherical or spheroidal surface. The plane containing this circle is parallel to the Equator and therefore its circumference is called a **Parallel of Latitude** or just **Parallel**. Since the plane is parallel to the Equator, it cannot pass through the centre of the Earth, and therefore a parallel is a small circle. The radius of the parallel in latitude φ is easily determined from the right-angled triangle NFG: FG = r = R sin χ; and since $ = 90° - \varphi$, r = R cos φ.

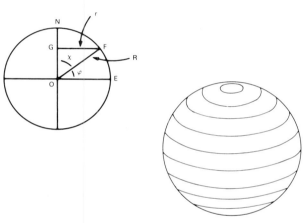

Fig. 2.26(a) Parallels

In Fig. 2.25 the locus of all points having the same longitude all lie within the same plane NPS, which traces a semi-circle on the surface of a sphere, or a semi-ellipse upon a spheroid. Since the plane passes through the centre of the Earth, it is the arc of a great circle and is known as a **Meridian**. Since the circumference of NSP intersects NGS at the two poles N and S, all meridans intersect at the poles.

Fig. 2.26(b) Meridians

Since the plane of the Equator is perpendicular to the axis NOS, it follows that all meridians intersect the Equator at right angles. Moreover, since all parallels of latitude are parallel to the Equator, **all parallels and all meridians intersect at right angles** on the curved surface of the sphere or spheroid. The geographical poles are two exceptional points at which all meridians intersect one another.

Fig. 2.26(c) Intersection of parallels and meridians

2.5 Plane co-ordinate systems

In order to plot the mathematical framework of a map it is desirable to use some kind of plane co-ordinate system. There are two simple systems in common use: plane polar co-ordinates; plane rectangular Cartesian co-ordinates. Both of these systems are commonly used in the study of the theory of map projections, but for the practical work of plotting a map, Cartesian co-ordinates are almost always used.

We study the meaning of the following key words: polar co-ordinates; plane rectangular Cartesian co-ordinates; origin; radius vector; vectorial angle; ordinate; abscissa; Easting; Northing; True North; Grid North; false origin.

2.5.1 Plane polar co-ordinates

The point O is selected as the **origin** from which measurements are to be made. The line OA is chosen as the **axis** or **initial line**. The position of any point P may be referred to this origin and axis by means of the **radius vector**, or straight line distance OP = r, and the **vectorial angle**, or angle AOP = θ. The position of P is recorded by the two quantities (r, θ).

We study the meaning of the following key
words:
- Polar coordinates
- Plane rectangular Cartesian coordinates
- Origin
- Radius vector
- Vectorial angle
- Ordinate
- Abscissa
- Easting
- Northing
- True North
- Grid North
- False origin

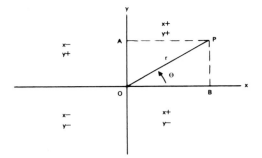

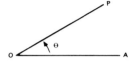

Fig. 2.27

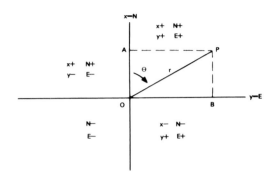

Fig. 2.28

It should be noted that in mathematics the angle
θ is measured counter clockwise from the initial
line. In surveying, navigation and cartography,
angles are measured in a clockwise direction.
Many of the instruments used in cartography use
this convention. The use of the clockwise convention introduces complications to the study of the
theory of map projections and in the practical
applications of computer programming.

2.5.2   Plane rectangular Cartesian co-ordinates

These may also be called **Cartesian co-ordinates**
or just **rectangular co-ordinates**. The position of
any point P may be referred to the origin, O, of the
system by means of two linear measurements AP
= OB and BP = OA, made along two perpendicular
axes intersecting at the origin. We refer to the line
OB as the **abscissa** and to OA as the **ordinate**. The
customary mathematical convention is to call the
abscissa the **x-axis** and refer to linear distances
such as OB as x. The ordinate is called the **y-axis**
so that the distance ON = y. The position of P is
defined as (x,y). Because of the convention that
angles are measured clockwise in cartography,
and the resulting difficulties mentioned in subsection 2.5.1, it is theoretically convenient to
transpose the meaning of the two axes so that the
ordinate is labelled the x-axis and the abscissa
becomes y. The two diagrams show that, in each
case, the angle is measured from the x-axis in a
positive direction.

Because of the confusion which may arise from
using the letters x and y for both systems, it is
better to use the letters E (for Easting) and N (for
Northing) to denote the system, which is more
convenient for the surveyor and cartographer.
These terms are used in making a **Grid Reference**
on a map. For example, P has co-ordinates (E, N).

2.5.3   Transformation from polar into rectangular co-ordinates

This is an extremely common calculation which is
needed to convert certain map projections into a
form which is suitable for plotting. If O is the
common origin of both polar and Cartesian co-
ordinates and the initial line of the polar system
coincides with the ordinate of the Cartesian
system, then P = (r, θ) = (E, N).

From the right-angled triangles: OB = E = r sin
θ; OA = N = r cos θ.

If it is required to convert rectangular co-
ordinates into polar co-ordinates: $r = \sqrt{E^2 + N^2}$;
tan θ = E/N.

Many modern calculators have these transformations built in as standard sub-routines. If the
user does not have access to a calculator which
solves square roots directly, it is easier to use: r =
N sec θ = E cosec θ to obtain the radius vector.

32

2.5.4 The map grid as a particular kind of plane reference system

A grid was briefly described in 2.2.3. We now see that it is simply a plane Cartesian reference system which satisfies the following rules:

1. The origin of the grid is defined as a particular point on the Earth's surface. In the example, this is the point 49°N, 2°W.

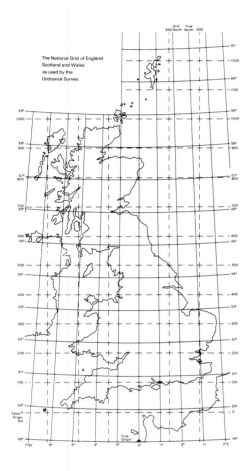

Fig. 2.29 The National Grid of England, Wales and Scotland as used by the Ordnance Survey

2. The orientation of the ordinate is also carefully defined. Usually this is the direction of the meridian passing through the origin. In the example, this is the meridian 2°W.
3. Consequently, the abscissa measures distances East or West of the origin. Since the sign convention indicates positive co-ordinates to the right of the origin, we use the term Easting for these measurements. The corresponding measurements along the ordinate are Northings, because this is the positive side of the origin.
4. The unit of measurement is usually the metre.
5. In order to overcome the disadvantage of having any negative co-ordinates for points which lie South and West of the origin, it is usual to re-number the axes to ensure that all co-ordinates are positive. This is equivalent to creating a **False Origin** for the grid. This point is located beyond the south-west extremity of the country to be mapped on the grid and is the point E = 0, N = 0 on the map illustrated.

2.5.5 A summary of the differences between a grid and a graticule

Graticule
Units of measurement: graticule lines are for equal values of latitude and longitude.
Order of recording co-ordinates: **latitude** followed by **longitude**.
Indication of direction: **True North** on a map is the direction of the meridian at any point.

Grid
Units of measurement: grid lines are for equal values of linear distances, usually metres.
Order of recording co-ordinates: **Easting** followed by **Northing**.
Indication of direction: **Grid North** on a map is the direction of the grid lines, which are parallel to the ordinate. It coincides with True North only along the meridian passing through the true origin of the grid.

TECHNIQUES

2.6 Methods of plotting the geometrical framework of a map

A general indication of the ways in which the grid or graticule of a map is computed, plotted and drawn has been shown diagrammatically on page 21. However, the information which is needed on the map and the way in which it is done depends upon the scale and purpose of the map. We distinguish three main possibilities, which are shown in Table 2.3.

From this flow diagram we see that there are two plotting stages, one calculation stage and the fair drawing stage. We consider each of these in detail in the sections which follow.

2.6.1 Instruments for plotting

The simple instruments which may be used for plotting are:
Marking aids: plotting needle in holder; pencil.
Guides for drawing: straight edge; compasses; set squares; splines and curves.
Measuring instruments: steel scale; protractor; beam compass; station pointer; precision dividers; bowspring dividers.

Table 2.3

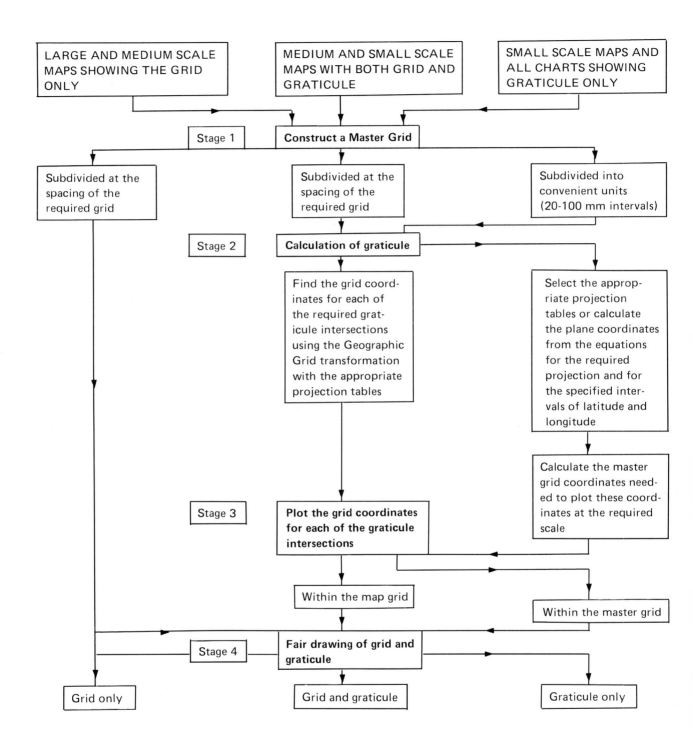

Fig. 2.30

Because these are described and illustrated in Section 2.7, there is no need to comment about them in this section. Special instruments used for plotting co-ordinates—the co-ordinatographs and the master grid template—are described in 2.6.6 and 2.6.7.

2.6.2 Limitations of the graphical procedures

Only four geometrical operations can be undertaken using simple drawing instruments:
- —To **draw** a straight line; ①
- —To **measure** the length of a straight line; ②
- —To **draw** a circular arc; ③
- —To **measure** or **construct** an angle. ④

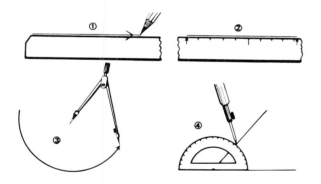

Fig. 2.31

If it is necessary to plot and draw a curve which is more complicated than the arc of a circle, the draftsman must plot a series of points which lie on that curve using some of these operations. The line representing the curve is traced through the plotted points, using a **spline** or similar device to act as the guide for drawing.

2.6.3 Location of a point on a plane

The minimum data which are needed to locate an **unknown** point on a plane with respect to one or more points of **known** position are shown below.

In the diagram the position of A and the direction AN are already known. The unknown point, P, may be plotted by means of the angle NAP and the distance AP.

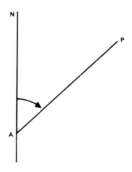

Fig. 2.32

2.6.3.1 Bearing and distance
By construction of one angle and one distance from **one** known point and **one** known direction through that point. This is the method of polar co-ordinates (2.5.1) and, in certain applications such as navigation, it is also known as **bearing and distance**.

2.6.3.2 Intersection
By construction of two angles at **two** points measured from the straight line joining them. This procedure is known as **intersection** in surveying. In the diagram the positions of A and B are known. Therefore, we must also know the distance AB and the direction of that side of the triangle, ABP. If we plot the angles ABP and BAP, this is sufficient to complete the triangle and therefore locate the position of P.

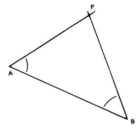

Fig. 2.33

2.6.3.3 Linear measurements
By construction of two linear distances from **two** known points. In the general case this is the geometrical construction of the triangle ABP where all three side lengths are known. A special application of the method is used for plotting Cartesian co-ordinates on axes of a grid as is illustrated in sub-section 2.5.2.

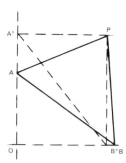

Fig. 2.34

In this diagram the co-ordinates of P are x = OB and y = OA. Since the side AB can be computed from $AB^2 = x^2 + y^2$, all the sides of the triangle APB are known.

2.6.3.4 Resection

By construction of three angles which have been measured at the unknown point, P. In surveying this is known as **resection**. In the diagram A, B and C are three known points. The three angles measured at P are α, β and γ. This method gives a unique solution for the position of P provided that the four points do not all lie on the circumference of the same circle. In that case there is no unique solution. This method is more commonly used to plot the detail of hydrographic surveys (using a special kind of protractor known as a Station Pointer) than in the other branches of cartography.

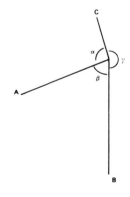

Fig. 2.35

2.6.4 Some geometrical fundamentals of plotting

The map manuscript, or **plotting sheet**, is a plane surface. Therefore, the definitions and theorems of plane, or Euclidean, geometry apply to the calculations and techniques used to plot points and draw lines: a point has position but no size; a line has length but no width.

In practice, we cannot locate a point which has no size, because this would be invisible. We cannot see a line with no width. Consequently, any legible graphical construction must contradict

these definitions. A point pricked with a fine needle represents a circle with a diameter of about 0.1 mm, a line scratched with a needle point also has a width of about 0.1 mm. These are the practical limits of what can be done in ordinary graphical work.

Most of the **Map Accuracy Standards** used by cartographic organisations to monitor the quality of their work refer to the positional accuracy of points of detail **which are measured from the grid or graticule of the map**. It follows that the geometrical framework of the map should be as free from error as can be practically accomplished.

2.6.4.1 Two fundamental principles

In order to achieve the exceedingly high accuracy which is therefore demanded, it is necessary to appreciate the nature of the errors of measurement and plotting which are likely to occur. We state two fundamental principles:

1. It is always desirable to make an **independent check** of each step in graphical construction. The methods described in 2.6.3.1–2.6.3.4 represent the **minimum** data which are needed to locate an unknown point on the plotting sheet. If any of the measurements are incorrectly made or transferred, the point will be plotted in the wrong place. We therefore make some additional measurements to check the accuracy of the work and confirm that the point has been located correctly.

2. It is always desirable to **interpolate** rather than **extrapolate**. This may be described as the principle that it is always desirable **to work from the whole to the part**. For example, it is better to measure and plot a line of length 700 mm and then subdivide this into 10 equal parts than to construct the same line by stepping off 10 measurements each of length 70 mm (see also 2.7.2.1 for the detailed analysis of this principle).

2.6.5 The master grid

The most convenient form of plotting sheet for construction of the grid or graticule of a new map is a **Master Grid** which is a plane Cartesian co-ordinate system plotted or reproduced on the drafting material to be used for compilation. Nowadays this is likely to be matt-surfaced polyester plastic, but other materials, such as enamelled zinc, laminated card or even coated glass, have been used for this purpose.

The grid lines should be plotted at equidistant intervals. For large and medium scale maps showing a grid, the spacing of these lines should be those required on the finished map. For maps without a grid, the spacing of the master grid lines should be about 20–50 mm if the graticule is to be plotted by manual methods. A wider spacing

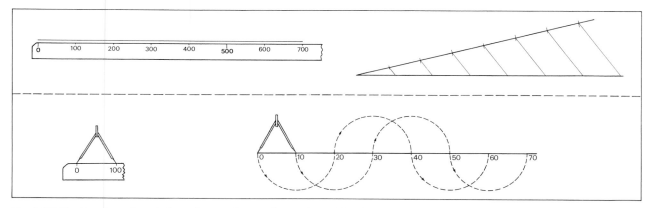

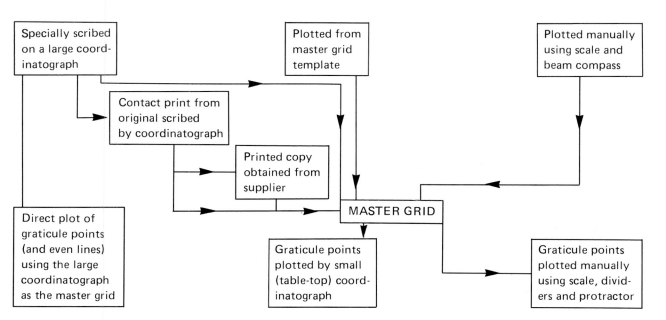

Fig. 2.36 Working from the whole to the part (upper diagram); working from the part to the whole (lower diagram)

(up to 100 mm) is adequate if the graticule points are to be plotted by a small co-ordinatograph.

Once the graticule has been plotted, the master grid lines are no longer required. If these have been drawn or reproduced on the plotting sheet in blue, they will disappear during subsequent photographic processing. It is important that the lines of the master grid should be as fine as is compatible with legibility. Lines of width 0.1−0.15 mm are quite thick enough.

2.6.5.1 The procedures for preparing and using the master grid
These are summarised in Table 2.4.

2.6.6 The co-ordinatograph

This is an instrument which is specifically intended to plot or measure the plane co-ordinates of points. The majority of these instruments, including all the large, free-standing models used for cartographic work, operate in Cartesian co-ordinates.

A large co-ordinatograph has a working range of 1 000 × 1 000 mm or more. It comprises a glass-topped table over which the axes of the instrument are permanently mounted. The instrument is large and heavy and must be erected on a stable floor in a room which is not subject to large temperature changes. The co-ordinate axes are represented by

Table 2.4

two massive steel bars. One of these is rigidly fixed to the table; the other can be moved in a direction perpendicular to the length of the fixed bar. A plotting device is attached to the moving axis and this too can be moved along the bar. Both of the movements operate scales through a rack-and-pinion system for instruments specifically intended for cartographic use, or through lead screws on the co-ordinatographs attached to photogrammetric plotters. Usually the scales are read under magnification and verniers permit measurement along each axis to the nearest 0.01 mm. The plotting device will hold a variety of interchangeable components, such as point makers, reading microscopes, drawing and scribing tools. Some co-ordinatographs also have a beam compass attachment which can be fixed to the plotting device and is used to draw or scribe circular arcs.

Fig. 2.38

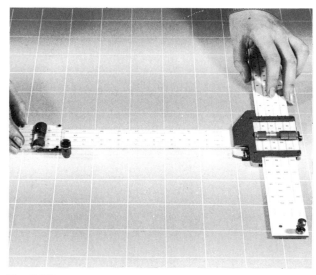

Fig. 2.37

A large co-ordinatograph is the ideal instrument for plotting a grid or graticule. If it is available for long periods of work, there is no need to prepare a special master grid because the instrument itself acts as the grid within which the graticule and other points may be plotted.

However, manual setting of the verniers takes rather a long time and therefore preparation of a complicated graticule is slow. If the cartographer has only limited access to a co-ordinatograph (such as one which is normally used for photogrammetric plotting), it is more economical to scribe a suitable master grid by co-ordinatograph and then use this copy as the negative to make additional grids as these are required.

The small size co-ordinatographs are essentially scaled-down versions of the large instrument to give a working range of only 250 × 250 mm or 500 × 500 mm. These are portable instruments

to be placed over the plotting sheet on a drawing board or tracing table. Most of them are more simple and crude than the large co-ordinatograph.

For example, the axes may slide against one another without any gears or roller bearings. The scale can only be read to 0.1 mm or thereabouts.

2.6.7 The master grid template

This is a sheet of metal, measuring about 800 × 600 mm, in which have been drilled a regular network of holes.

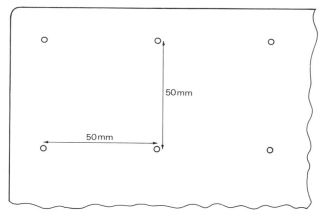

Fig. 2.39

The centres of these holes have all been located with great accuracy and the diameters of them are all identical. A small pricking tool (Fig. 2.40) is used to mark the points. The external diameter of the shaft which contains the pricking needle is only a few micrometres less than that of the holes, so that the instrument fits snugly without sideways movement. The spacing of the holes is usually either 50 or 100 mm, forming a square grid pattern.

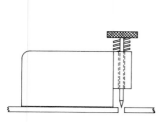

Fig. 2.40

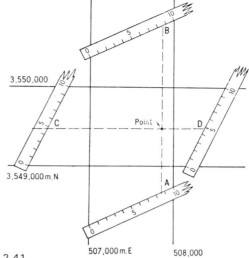

Fig. 2.41

In order to prepare a master grid by template, the plotting sheet is mounted on a table or drawing board. The template is laid over the plotting sheet and the individual points are pricked with the tool. It is advisable to mark each point immediately after it has been plotted and **before the template is removed**. This can be done by drawing a small circle within each hole using a pencil with a long, sharp point. The grid is completed by joining the marked points by straight edge. This is less accurate than ruling the lines by co-ordinatograph at the same time as the points are plotted, but in all other respects the master grid template is as accurate as the large size co-ordinatograph. Moreover, it is a much quicker way of preparing a 50 or 100 mm grid, because there are no scales to be read or set.

The combination of the master grid template with a table-top co-ordinatograph is a particularly efficient way of plotting a graticule, and it is much cheaper to install than a large co-ordinatograph. If a master grid prepared by template is to be used for graphical plotting, it is advisable to subdivide this further (e.g. with lines at 20-mm intervals). This is done graphically.

2.6.8 Plotting points within the master grid

Because the master grid is a Cartesian system, the individual points representing graticule intersections are plotted from linear distances.

There are two techniques in common use: using dividers; using a short steel scale (an Engineer Scale). The second method is a practical modification of the well-known geometrical method of subdividing a line into equal parts.

The diagram illustrates how a point with grid co-ordinates 507864.3m E, 3549487.7 m N may be plotted by placing a scale in each of the positions A, B, C and D.

2.6.9 Automation in plotting

The graphic output from a digital computer suitable for cartographic purposes is by **Flat-Bed Plotter**. This is similar in most respects to a large co-ordinatograph, with the important difference that scales do not have to be set manually or read by eye and the plotting device can be lowered or raised automatically. All the movements are controlled by servo-motors or electromagnetic devices in obedience to signals generated in computation.

Since equipment of this sort is described more fully in Volume 2, we need not devote further space to the description or illustration of it here. It will suffice to state that the flat-bed plotter will plot a grid or graticule and the neat lines of a map without manual intervention. Moreover, there is no constraint upon the complexity of the curves which have to be drawn. A manually operated co-ordinatograph can only be used to draw straight lines or, with a beam compass attachment, circular arcs.

2.6.10 Stages in graphical construction of a grid

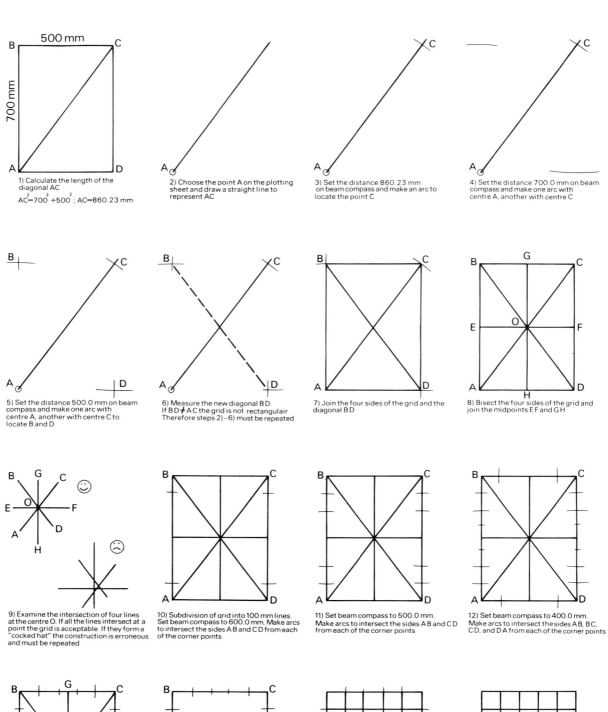

1) Calculate the length of the diagonal AC
$$AC^2 = 700^2 + 500^2 ; AC = 860.23 \text{ mm}$$

2) Choose the point A on the plotting sheet and draw a straight line to represent AC

3) Set the distance 860.23 mm on beam compass and make an arc to locate the point C

4) Set the distance 700.0 mm on beam compass and make one arc with centre A, another with centre C

5) Set the distance 500.0 mm on beam compass and make one arc with centre A, another with centre C to locate B and D

6) Measure the new diagonal B D. If B D ≠ A C the grid is not rectangulair Therefore steps 2)–6) must be repeated

7) Join the four sides of the grid and the diagonal B D

8) Bisect the four sides of the grid and join the midpoints E F and G H

9) Examine the intersection of four lines at the centre O. If all the lines intersect at a point the grid is acceptable. If they form a "cocked hat" the construction is erroneous and must be repeated

10) Subdivision of grid into 100 mm lines. Set beam compass to 600.0 mm. Make arcs to intersect the sides A B and C D from each of the corner points

11) Set beam compass to 500.0 mm. Make arcs to intersect the sides A B and C D from each of the corner points

12) Set beam compass to 400.0 mm. Make arcs to intersect the sides A B, B C, C D, and D A from each of the corner points

13) Set beam compass to 300.0 mm. Make arcs to intersect the sides B C and D A from each of the corner points

14 Remove the construction lines A C, B D, E F and G H

15) Join subdivisions along sides with straight lines

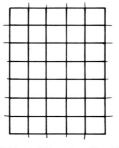
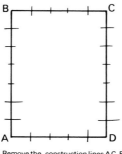
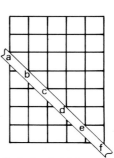

16) Check ruling of subdivisions by placing a straight edge diagonally across grid. The edge should pass through all the intersections a, b, c, d, e, and f

Fig. 2.42

40

2.7 The nature of the errors of measurement and plotting

In any science or technology concerned with measurements, the errors which will arise are usually classified as: gross errors; systematic errors; random errors. All of these can arise in cartographic practice.

2.7.1 Gross errors

These are mistakes which are made in reading a scale, copying the results of a measurement, or mishandling an instrument. For example, in cartography:

—**Incorrect scale reading**: It is required to set a beam compass against a steel scale to plot a distance 567.8 mm, but, in concentrating upon estimation of 0.8 mm between the sub-divisions of the scale, the setting of the compass is made 566.8 mm, 557.8 mm or even 667.8 mm.

—**Transposition of numbers in copying**: In making a list of the measurements to be plotted, the distance 567.8 has been written as 576.8 or some other sequence of these numbers.

—**Accidental alteration in setting of instruments**: The separation of the points of dividers may be accidentally changed as the measurement is transferred from the scale to the plotting sheet.

Each of these is an example of a mistake which arises from carelessness or inattention, but which can be recognised and rectified by repetition of the measurement or plotting process. Usually the accidental displacement of the setting of a pair of dividers can be recognised if comparison is made with the scale for a second time after plotting the distance. This is because the displacement is likely to be irregular and repetition of the scale comparison **looks wrong** when it is done a second time. On the other hand, it is easy to repeat a gross scale error in reading the scale of a co-ordinatograph or checking a beam compass distance, simply because the scale distance **looks right** when it is done by the same person, using the same instrument in the same way. Although the gross error would probably be detected at once if any of the variables operator/instrument/method were changed, this is inconvenient in production cartography because it effectively doubles the time needed to do the work. It is therefore preferable to use a system of checking which is independent but less time-consuming. This comprises making and plotting at least one more measurement than is theoretically necessary.

We have already noted in 2.6.3 that the methods described make use of minimum data to locate an unknown point on a plane. They are minimum data in the sense that when correctly measured the point is correctly located, but they all suffer from the disadvantage that one gross error in measurement or plotting will put the unknown point in the wrong place on the map. We overcome the risk of plotting gross errors by using more than the minimum data.

For example, intersection might be extended to locate a point with respect to three known points instead of only two. Similarly, four grid distances should be calculated and plotted instead of two.

Although these additional measurements are **redundant** in the sense that the use of them does not affect the location of a point which has already been correctly plotted using the minimum data, they serve as an important check upon the presence of gross errors. In other words: **The first fundamental principle of plotting is intended to eliminate any gross errors.**

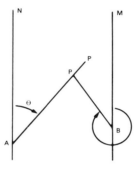

Fig. 2.43(a) The use of redundant measurements to serve as a check against gross errors. The point P is to be plotted by polar co-ordinates from A by means of the angle θ and the radius vector AP. The angle is correctly measured but a gross error in measurement of the distance AP locates the point at P'. A check measurement using the angle MBP and the distance BP shows that P' is incorrect

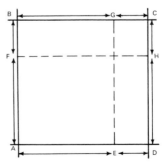

Fig. 2.43(b) The use of redundant measurements to serve as a check against gross errors in plotting a point by rectangular co-ordinates. The measurements AE = BG = x and AF = DH = y are sufficient to locate P within the grid. The independent measurements GC = ED = x' and FB = HC = y' are an extra check on the plot

41

2.7.2 Systematic errors

These are small errors, but they are cumulative and therefore grow into larger errors if they are repeated and not checked. In cartography they usually arise from employing an **unsuitable method** of doing the work rather than from any carelessness in making measurements or handling instruments. We give two examples.

2.7.2.1 Systematic errors in plotting distances
As illustrated in 2.6.4.1, a common, but unsatisfactory, method of dividing a line into equal parts is to set a pair of dividers to a separation corresponding to the single division. The dividers are set at one end of the line and 'walked' along the line, moving each point in turn.

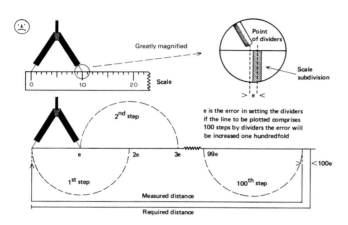

Fig. 2.44

If this method is used to subdivide a line which has already been plotted, it is nearly always possible to detect a discrepancy in the position of the end point located by the two methods. This arises from the fact that it is seldom possible to set the dividers **exactly** by comparison with a scale. There will be a very small error, e, between the position of the dividers point and the correct position on the scale. This error is repeated with every step made by dividers. Suppose that the length of the line to be plotted is 700 mm and this is done in 10 steps of 70 mm each. If the cartographer can estimate distance on the scale to the nearest 0.1 mm, any setting of the dividers between 69.95 mm and 70.05 mm would appear to be correct. If it is 69.95 mm, e = − 0.05 mm, which is a negligible error. But in a line comprising 10 steps, 10 e = − 0.5 mm and the plotted length of the line would be 699.5 mm. In other words, a small systematic error has grown into a discrepancy which is easily measured.

2.7.2.2 Systematic errors in plotting angles
Suppose that it is required to plot the angle ABQ in order to locate point Q. The usual method is to place a protractor over the point B, align the zero mark with the line BA and prick the point P against the required angle given by the scale. The construction is completed by drawing the straight line BP and locating Q upon it. If BQ is less than BP, this construction is adequate. This is because we are working from the whole (the radius of the protractor) to the part (the short line BQ). On the other hand, if the required point lies further away than the edge of the protractor, it would be necessary to extend the line BP to Q.

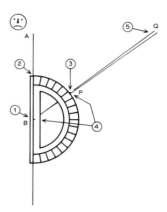

Fig. 2.45 Erroneous construction of an angle by protractor

The following errors may arise in making the measurement and construction:
- a small error is made in placing the centre of the protractor over the point B; ①
- there is a small error in alignment of the zero mark with the line BA; ②
- the point P is incorrectly marked owing to a small error in reading the scale of the protractor; ③
- the straight line BP does not pass exactly through these points. ④

Any one of these errors may have been made, but each of them is so small that they cannot be easily detected by eye. But in drawing the line BQ these errors are magnified in proportion BQ:BP ⑤ . Thus, if the radius of the protractor is 100 mm and the distance BQ is 1 000 m, an error e = 0.1 mm in the location of P will be increased to the error (1 000/100) e = 1.0 mm in the location of Q.

In order to avoid this kind of systematic error it is necessary to work from the whole to the part, and construct the angle ABQ with a beam compass and scale using the calculated **chord distance** corresponding to the required angle.

It follows that: **The second fundamental principle of plotting relates to the elimination of systematic errors.**

2.7.3 Random errors

These are the errors which remain after the gross and systematic errors have been eliminated. Probably the most important single cause of random errors in cartography arises from the uncertainty which arises in estimating the distance between the engraved subdivisions of a scale. It is usual for a draftsman to attempt mental subdivision of the spaces between the marks into tenths. Thus a scale which is marked at intervals of 1.0 mm should be divided by estimation into units of 0.1 mm. However, this leaves room for some uncertainty. An estimate of 0.6 must serve for all measurements between 0.55 and 0.65; an estimate of 0.7 represents the range 0.65−0.75 and so on. The limit of reliable estimation of any scale is known as its **resolution**. In this example, resolution is 0.1 mm.

Consider a line of length 152.45 mm. If scale resolution were the sole source of error, a series of repeated measurements of this line would give a series of readings one-half of which would be 152.4 mm and one-half of which would be 152.5 mm. Therefore, a graph showing these measurements would be rectangular in shape. It is known as a **Rectangular Frequency Distribution**.

Superimposed upon these variations are some which are less well-defined. One of these is the **personal equation** of the draftsman which is his inability to make reliable mental subdivision of a scale. There is a common tendency to give emphasis to some subdivisions—usually 0.2, 0.3, 0.7 and 0.8 at the expense of 0.4, 0.5 and 0.6. Consequently, the spread of readings may exceed 0.2 units of the space between scale subdivisions. Other causes of error include the variations in illumination of the scale and plotting sheet and the effects of this upon the draftsman's eyesight.

Slight flexibility in the rigid arms of instruments and unequal penetration of dividers points into the plotting sheet also contribute small errors.

The combined effect of all these sources of error is that:

—small errors are more common than large errors;

—positive and negative errors are equally likely to occur;

—if a single measurement is repeated many times, the frequency with which a particular result occurs has a characteristic distribution. If this is plotted as a graph, it has a bell-shaped outline and is known as the **Normal Distribution** or the **Normal Curve**.

2.8 Calculations involved in preparing a grid or graticule

It was seen in 2.6 that the construction of the geometrical framework of a map requires some calculations. Usually these are quite simple, for even the data required in the co-ordinate transformations from the spheroid are tabulated in **Projection Tables** in a way which requires com-

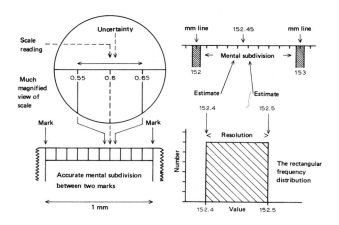

Fig. 2.46 The rectangular frequency distribution

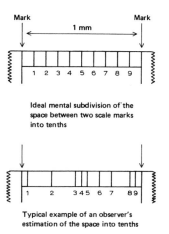

Fig. 2.47

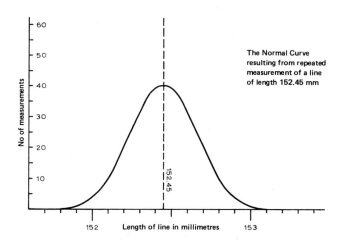

Fig. 2.48 The normal curve resulting from repeated measurements of a line of length 152.45 mm

paratively little additional calculation, and certainly does not require a knowledge of the complicated equations from which they have been derived.

In almost every case the calculations can be made by desk or pocket calculator and many of them can also be solved by slide rule. However, an important practical feature of cartographic work is the repetitive nature of many computations. In order to plot a map projection it may be necessary to repeat the solution for every point to be plotted. Since there may be several hundred of these, the work of calculation may be laborious and time-consuming. There is consequently a strong case to be made for solving the equations by digital computer, because a computer is particularly well suited to deal with repetitive calculations.

The cost of execution of the work by using a modern computer is small compared with the labour costs which arise from making and checking a large number of individual calculations by hand.

2.8.1 Large- and medium-scale maps showing a grid only

The only calculation to be made is that required to convert from the specified grid spacing in kilometres to the corresponding distance in millimetres on the plotting sheet. Usually the spacing of the grid lines is either 0.1, 1.0 or 10.0 km (depending upon the scale of the map). We use the scale conversion formula described in (1) of 2.2.1:
Specification:
   1. Scale of map: 1/2 500.
   2. Grid interval: 0.1 km.
Solution:
   Grid spacing: $\dfrac{0.1 \times 1\,000\,000}{2\,500} = 40.0$ mm.

Table 2.5 shows the spacing of the grid lines (in mm) for some of the commonly used map scales, and the grid intervals which are usually specified.

Table 2.5

Scale of map	Grid spacing 0.1 km	Grid spacing 1.0 km
1/500	200.0	—
1/1 000	100.0	—
1/1 250	80.0	—
1/2 500	40.0	—
1/5 000	20.0	200.0
1/10 000	—	100.0
1/12 500	—	80.0
1/25 000	—	40.0
1/50 000	—	20.0
1/100 000	—	10.0

2.8.2 Medium- and small-scale maps showing both grid and graticule

It is essential that the two methods of referencing

position should be mutually compatible. Therefore, it is necessary to plot the grid first, calculate these grid co-ordinates for each graticule intersection, and finally plot these within the grid.

Usually this kind of map is based upon the projection system adopted by a national survey for all topographical mapping. The projection is most likely to be a version of the **Transverse Mercator Projection**, **Lambert Conformal Conical Projection** or the **Polyconic Projection**, having been specified by that survey. Most national surveys publish projection tables and use must be made of these to determine the required grid co-ordinates. The transformation which is needed is usually called **Geographical-to-Grid**, which finds the (E, N) co-ordinates of a point from the known (φ, λ) co-ordinates.

We do not give a numerical example, because the way in which the calculations are organised depends upon the tables in use. Even for a uniform system like the UTM, the work may be tabulated in different ways.

Having obtained the grid co-ordinates of a graticule point there still remains the calculation needed to convert from metres on the ground into millimetres on the plotting sheet. The method is similar to that of example 1 in 2.2.1:

Specification:
  —scale of map 1/250 000;
  —geographical co-ordinates 10°N, 30°E;
  —UTM grid, zone 35, Clarke 1880 spheroid;
  —grid co-ordinates of this point: from UTM tables these are found to be 828 935.1 m E, 1 106 808.7 m N;
  —this point is to be plotted with reference to the grid intersection 820 000.0 m E, 1 100 000.0 m N.

Solution:
  Difference in Eastings: 828 935.1 − 820 000.0 = 8 935.1 m;
  Map distance:
$$\frac{8\,935.1 \times 1\,000}{250\,000} = 35.74 \text{ mm (E)};$$
  Difference in Northings: 1 106 808.7 − 1 100 000.0 = 6 808.7 m;
  Map distance:
$$\frac{6\,808.7 \times 1\,000}{250\,000} = 27.23 \text{ mm (N)}.$$

The values 35.74 and 27.23 mm represent the minimum data needed to plot this graticule point within the grid. Following the advice in 2.7.1 we should also check the distances from the grid point 830 000.0 m E and 1 110 000.0 N.

Since the grid separation at 1/250 000 is 10 km (40.0 mm on the plotting sheet), the required measurements are 40.0 − 35.74 = 4.26 mm, and 40.0 − 27.23 = 12.77 mm, respectively.

2.8.3 Small-scale maps and charts with a graticule only

In this case the lines of the master grid are not intended to appear on the finished map and they merely serve as a guide for accurate location of the graticule intersections. Since maps which show a graticule only are usually at a scale smaller than 1/500 000, the assumption that the Earth is a perfect sphere will be adequate for all the calculations. Some navigation charts are based upon the spheroidal assumption. These calculations are not described in this Manual.

In the simplest case there are two stages in the transformation:
- Calculation of the projection co-ordinates from the equations which define the map projection to be used.
- Conversion of these into the master grid co-ordinates which are used to plot the projection at the required scale.

2.8.3.1 Solution of projection equations

In some of the standard textbooks on map projections the descriptions include tables of (x, y) or (r, θ) co-ordinates. If these are available, there is no need to calculate the projection co-ordinates. Often, however, the required projection differs from the information which is available. For example, it may be desirable to use a particular **aspect** (see 2.12), or a different **standard parallel** (see 2.14.2), so that the tabulated information has no use.

A typical example of the calculations which are required for a projection with simple equations is: Specification: Scale 1/40 000 000.

Sinusoidal projection (equations see 2.14.4): $X = \varphi$; $Y = \lambda . \cos \varphi$.

We note that the axes convention is that $+x$ is towards the North and $+y$ is towards the East.

Origin of projection co-ordinates: $\varphi_0 = 0°$, $\lambda_0 = 0°$.

Required graticule point: $\varphi = 30°N$, $\lambda = 30°E$.
Solution:

From tables to convert from degrees into radians: $\varphi = \lambda = 30° = 0.5236$ radians.

From tables of trigonometric functions: $\cos 30° = 0.8660$.

The required projection co-ordinates are: $Y = 0.5236 \times 0.8660 = +0.4535$; $x = +0.5236$. Since the point lies North and East of the origin, both x and y are positive.

2.8.3.2 Conversion to master grid co-ordinates

The projection co-ordinates are derived for a **sphere with unit radius**, i.e. $R = 1.0$. It is necessary to convert these values into the dimensions needed to plot the map on a millimetre master grid at the required scale.

In 2.3.4 we found that the radius of the Earth is 6 371.1 km. Applying the scale conversion formula to this distance, for the scale 1/40 000 000:

$$\frac{1}{40\ 000\ 000} = \frac{R'}{6\ 371\ 100}$$

or $R' = 0.1592$ m $= 159.2$ mm.

We multiply all the (x, y) projection co-ordinates by the constant R' in order to find the x' and y' master grid co-ordinates to be plotted.

In the example where $Y = +0.4335$, $X = +0.5236$, $Y' = 0.4535 \times 159.2 = +72.2$ mm, $X' = 0.5236 \times 159.2 = +83.3$ mm; these are the values to be plotted on the master grid.

2.8.3.3 Variations in procedure

The examples given in 2.8.3.1 and 2.8.3.2 illustrate the kind of calculations which are needed to plot any graticule point. There are, however, a number of short cuts which can be used to simplify calculation and plotting of the points needed to construct the entire graticule.

1. In the case of the cylindrical projections (2.14.1), the normal aspect graticule is composed of a rectangular network of straight lines. Because all the parallels and meridians can be drawn by ruler through pairs of points at the ends of each line, it is sufficient to calculate the co-ordinates for only these terminal points and then draw the graticule.

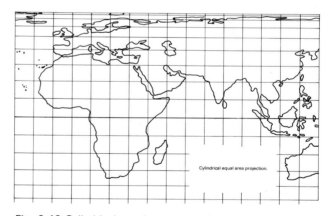

Fig. 2.49 Cylindrical equal-area projection

2. Many map projections are symmetrical about one or two axes. In these cases it is sufficient to calculate the (x, y) co-ordinates for only one-half (usually East of the central meridian) or for one-quarter (East of the central meridian and North of the Equator) of the whole map. The intersections in the other hemisphere or other quadrants can be found by applying the sign convention, used for Cartesian co-ordinates, to the (x, y) co-ordinates. In the example of the Hammer-Aitoff projection which is illustrated, we may compute (x, y) co-ordinates for the point 60°N, 60°E as:

Point		Y	X
60°N	60°E	+0.5907	+1.0231

From the symmetry about the Equator:

60°S	60°E	+0.5907	−1.0231

From the symmetry about the central meridian:

60°N	60°W	−0.5907	+1.0231
60°S	60°W	−0.5907	−1.0231

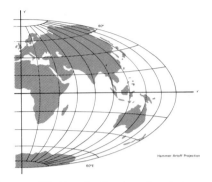

Fig. 2.50 Hammer-Aitoff projection (one hemisphere only)

3. If there are no axes of symmetry which coincide with the graticule, it is necessary to compute the co-ordinates of each point separately. For example, in order to plot this oblique aspect of the sinusoidal projection, it would be necessary to compute the projection co-ordinates for 227 individual points on a world map, having 15° intervals of latitude and longitude.

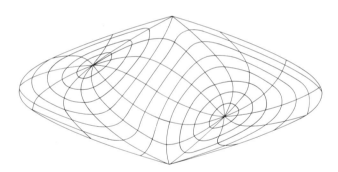

Fig. 2.51 Oblique aspect sinusoidal projection. No axes of symmetry

2.8.3.4 Complications in procedure

Many map projections are best described in polar (r, θ) co-ordinates. In order to plot these on a Cartesian master grid it is necessary to use the transformation described in 2.5.3. In many cases the radius vector is a meridian and the vectorial angle is related to longitude.

Therefore one of the axes of the master grid is made to coincide with a meridian (λ_0), and the vectorial angle is therefore a function of the difference in longitude, $\delta\lambda = \lambda_0 - \lambda$, for any graticule point on the meridian λ.

Specification:
Azimuthal equal-area projection (2.14.3) defined by the polar co-ordinates: $r = 2 \sin \frac{x}{2}$; $\theta = \lambda$.

Solution:
Make the prime meridian (λ_0) coincide with the abscissa of the master grid. Then, from 2.5.3:
$Y = 2 \sin \frac{x}{2} \sin \delta\lambda$; $X = 2 \sin \frac{x}{2} \cos \delta\lambda$.
Numerical values for x and y and finally x' and y' may be determined using the procedure described in 2.8.3.1 and 2.8.3.2.

The **aspects** of a map projection are defined and illustrated in 2.12. Projection co-ordinates are usually derived for the **normal aspect**, or simplest version of the graticule, where φ and λ occur in the projection equations. In order to calculate the co-ordinates for a map projection in a different aspect it is necessary to make an additional transformation *before* computing the projection co-ordinates.

—We start from the knowledge of the origin of the projection. This is the point with geographical co-ordinates φ_0 and λ_0.

—We compute the **bearing and distance co-ordinates** (z, α) of each required point (φ, λ) using the methods of spherical trigonometry. Using the notation which we have introduced for latitude and longitude, the two equations are:
$\cos z = \sin \Phi_0 \sin \Phi + \cos \Phi_0 \cos \Phi \cos \delta\lambda$;
$\sin \alpha = \cos \varphi \sin \delta\lambda \csc z$.

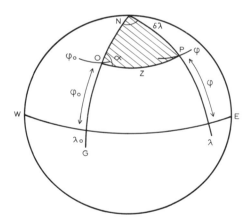

Fig. 2.52 Relationship between (φ, λ) geographical co-ordinates and (z, α) bearing and distance co-ordinates of the point P measured from the point O (φ_0, λ_0) as origin

As before, $\delta\lambda = \lambda - \lambda_0$, and is the difference in longitude between the point and the origin.

Expressed in bearing and distance co-ordinates, the projection co-ordinates for the azimuthal equal-area projection are: $r = 2 \sin \frac{z}{2}$, $\theta = \alpha$; or $Y = 2 \sin \frac{z}{2}$, $\sin \alpha$; $X = 2 \sin \frac{z}{2}$, $\cos \alpha$.

Tables are available to make the transformation from (φ, λ) into (z, α) for intervals of 5° of latitude and longitude, and for values of every 5° of latitude. Use of these much reduces the amount of computation needed.*

*Wagner, K. H., 1949. *Kartographisches Netzentwürfe*, Bibliographisches Institut, Leipzig, 262 pp.

46

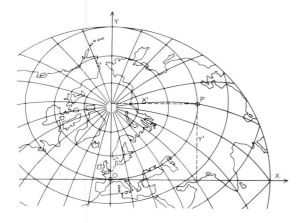

Fig. 2.53 The master grid co-ordinates (x', y') of the point P in an oblique aspect map projection

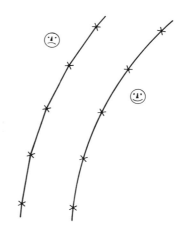

Fig. 2.54

2.9   Fair drawing of the graticule

It was stated in 2.6.2 that conventional drafting instruments can only be used to draw straight lines or circular arcs. At the fair drawing stage, a graticule which is entirely formed of these lines can be completed with ruler and compass. Alternatively, this kind of graticule can be drawn or scribed on a manually operated co-ordinatograph. Nevertheless, some difficulties arise in using wholly geometrical methods. For example, it is difficult to draw the arc of a circle having a large radius (1 000 mm or more) with requisite precision by beam compass. Often arcs with large radii cannot be drawn because the centre of the circle lies beyond the plotting sheet, or even beyond the edge of the drawing table.

In general, therefore, any graticule which has curved parallels or meridians must be drawn with the aid of **curves** or **splines** to rule continuous lines through all the points plotted along a single parallel or meridian.

These curves are intended to satisfy certain mathematical functions which must apply, not only at the points which have been plotted, but also at all intervening points along each curve. In other words, the smooth curve drawn through the plotted points has mathematical significance, and it will not suffice to join the points in an arbitrary fashion. For example, a curved meridian should not be drawn as a series of short straight lines between the plotted points.

An important aid to construction of complicated graticules for small-scale maps is to plot more points than need be shown on the completed map. For example, if the specification for an atlas map calls for 4° separation of the parallels and meridians, drawing the graticule (and plotting of map detail) is facilitated if the initial plot of the graticule shows 1°-intervals between the lines.

2.9.1   Splines

These are the principal aid to ruling curved lines, especially at larger map scales where the radii of curvature of the meridians or parallels are large. The conventional spline is a wooden rod about 1 m in length. A set of splines comprises fifteen or twenty such rods having different thickness and section. Some splines taper towards the ends and others are thinner in the middle.

A spline is placed upon the plotting sheet and set to lie through a succession of plotted points, using the flexibility of the wood to create a continuous curve. The rod is held under tension by means of **ship weights**. In order to draw a continuous curve there should be no weights along the outer edge of the spline where these would interrupt the drawing, but to create tension there must always be two weights on this side of the rod near the ends. Since we do not want the spline to move sideways during the drawing process, a series of additional weights are placed along the inside edge. Because the line must be drawn through the points which have been plotted, the position of the spline must be slightly offset in order to take account of the width of the ruling pen or the shaft of the scribing tool.

When the graticule is to be scribed, an alternative form of spline is easier to use. This is a strip

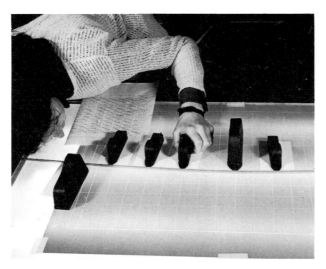

Fig. 2.55(a)

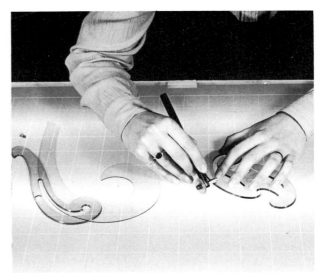

Fig. 2.56

Fig. 2.55(b)

of thick but flexible plastic (e.g. vinyl plastic about 2 mm thick) or a strip of stainless steel. This is placed on the plotting sheet in an edge-on position and steadied by means of ship weights at each end. The inner edge of the spline is packed with modelling clay, such as is used by children, and this provides a firm anchor for the plastic strip. After the line has been scribed, the modelling clay can be wiped from the scribe-coat without spoiling the quality of the manuscript for later work and for reproduction.

2.9.2 Curves

These are rigid plastic templates cut into a variety of curved patterns. They are also much used in other kinds of drafting, notably the ship-building industry, so that names like **Ship Curves** or **Yacht Curves** also describe them. A set of curves usually comprises 20−30 templates having different

sizes and shapes. The selection of a particular edge for ruling part of a graticule line is a matter of trial and error. Usually it is only possible to find a curve which will suit a short arc of the line to be drawn. It is therefore necessary to build up each parallel or meridian by means of a series of such arcs. It is quite difficult to achieve continuity of a smooth curve through all the plotted graticule intersections.

THEORY OF MAP PROJECTIONS

2.10   Introduction to the theory of map projections

A map projection is any systematic arrangement of meridians and parallels portraying the curved surface of the sphere or spheroid upon a plane.
We have seen, in 2.2, that the graticule, which is the easiest way of recognising and understanding the use of map projections, may have an infinite variety in the ways in which the parallels and meridians can be shown. The reason why an infinitely large number of map projections are theoretically possible is because there is no such thing as the 'perfect map projection'. It is mathematically impossible to represent the curved surface of the sphere or spheroid upon a plane which does not show some kind of **deformation** of the curved surface, which is equivalent to tearing or stretching.
From the hundreds of individual map projections which have been described, only about twenty are important in practical cartography—and most of those in small scale atlas maps. The cartographer who is concerned with topographical maps or navigation charts is likely to deal with only five or six different map projections.

48

2.10.1 Deformation of map projections

The idea of tearing can be recognised in any map projection which has an **edge**, representing the limit of the world map. This edge is a purely artificial boundary because the spherical surface is continuous in all directions and has no edges.

A map projection may have additional edges which result from the representation of the same parallel or the same meridian in more than one place on the map. Figure 2.57 shows one such example, where each of the parallels 15°, 45° and 75° North and South are shown twice and parts of the meridian 180° are shown in ten different places. If we imagine the Earth to have a skin like a fruit and that cuts have been made through this skin in the places indicated by this map, the curved surface of the skin can be laid almost perfectly flat upon a plane. Hence this kind of map projection has little deformation **within** the component strips.

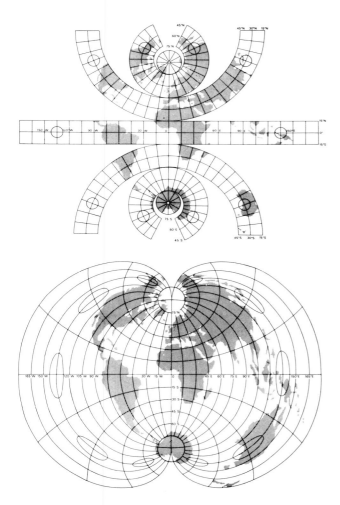

Fig. 2.57 World map on the polyconic projection (see 2.14.6, example 1). Note how the small circles on the diagram above have become ellipses on this map

However, it is inconvenient to show the continuous spherical surface interrupted by so many gaps. If it were desired to map the Earth continuously after this fashion, it would be necessary to stretch each part in a North—South direction until the intervening gaps have been filled. This can be done as illustrated, but clearly the process of stretching has altered the **scale** of the map in the North—South direction and the amount by which the scale has changed increases progressively from the centre of the map towards its eastern and western edges. In other words, **stretching involves changes in scale**.

2.10.2 Scale on a map projection

The elementary concept of scale was defined in 2.2.1. From this definition it is reasonable to make the following assumptions about distances measured on maps:

- The scale of the map is constant *for all distances*. Thus, if 40 mm corresponds to 1 km on a map of scale 1/25 000, we also assume that 80 mm corresponds to 2 km and that 20 mm represents 500 m.
- The scale of the map is constant in *all parts of the map*. In other words, a line of length 40 mm corresponds to 1 km on the ground whether this line be located near the middle of the map or near a corner or an edge of the sheet.
- The scale of the map is constant *for all directions on the map*. Thus 40 mm represents 1 km in the North—South direction, the East—West direction or in any intermediate direction.

None of these assumptions are correct.

Because some deformation must be present on the plane map, scale must vary from place to place and often in different directions at the same point. Examination of the two maps shown in Fig 2.57 shows that meridional stretching has had the effect of increasing scale in the direction of the meridians, but the separation between the meridians is the same on both maps. This is easy to detect upon a world map at a very small scale, but it cannot be seen or measured on a large-scale map. At the scale 1/25 000 used in the example, it is impossible to detect variation in scale with length of line, position or direction within the confines of the typical 1/25 000 scale map. This is because the variations within such a small area (100—200 km²) are so small that they cannot be measured. But this does not mean that they do not exist.

It follows that the elementary definition of scale in 2.2.1 is satisfactory for most kinds of map use on large- and medium-scale maps. It becomes less reliable in the study and use of maps of scale 1/1 000 000 or smaller.

49

2.10.3 Lines and points of zero distortion

Although it is impossible to preserve constant scale at all points and in all directions on a map, this can be done at certain points or along certain lines. Such a point is known as a **Point of Zero Distortion**. Similarly, we have one or more **Lines of Zero Distortion**. At these points, or along these lines, the scale of the map is equal to that on the surface of a **globe** representing the Earth's surface, which has been reduced by the same ratio as the scale stated on the map. For example, if 1 mm represents 1 km on the surface of a globe, the scale of the globe is 1/1 000 000.

2.10.3.1 Principal scale

If we now make a map of the globe in which the scale along certain lines (or at certain points) is also 1/1 000 000 we have created lines or points of zero distortion. We call this scale the **Principal Scale** and denote it numerically as 1.0.

2.10.3.2 Particular scales

At all other points on the map and, usually, in different directions from the same point, there is deformation which is equivalent to stretching of the surface of the globe. Consequently, the scales at these points and in these directions are larger or smaller than the principal scale. We call them **Particular Scale** and refer to them numerically by decimals. Thus a particular scale which is double that of the principal scale (1/500 000 in this example) is expressed as 2.0 and the particular scale which is one-half of the principal scale (or 1/2 000 000) is expressed as 0.5.

2.10.4 The ellipse of distortion

Where the particular scales vary with direction about a point it is easy to imagine that a very small circle centred at that point on the curved surface of the globe becomes deformed into an ellipse centred at the corresponding point on the map. This is called the **Ellipse of Distortion** and may be used as the basis for measurement of deformation of the map projection at that point. The distance from the centre of the ellipse to any point on its circumference is proportional to the particular scale in that direction. The two axes of the ellipse indicate the directions in which the particular scales are maximum and minimum for that point on the map.

In the diagrams, the point A has latitude φ and longitude λ on the globe. This is mapped on the plane as the point A' with co-ordinates x and y. The small circle on the spherical surface has a radius of 1.0, corresponding to the principal scale. The particular scales on the map which are commonly used in the study of deformation are:

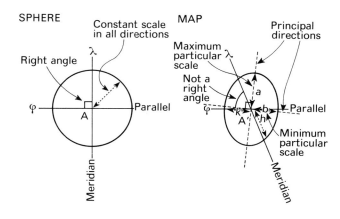

Fig. 2.58

—particular scale along the meridian = h;
—particular scale along the parallel = k;
—maximum particular scale = a;
—minimum particular scale = b.

These quantities may be calculated from the equations relating x and y to φ and λ for any given projection.*

2.10.4.1 Principal directions

The orientation of the two axes of the ellipse indicate two **Principal Directions**. Since the axes of an ellipse are perpendicular to one another, the principal directions of the ellipse of distortion are always at right angles. Moreover, this represents a special case where a right angle on the spherical surface is always represented by a right angle on the map. Thus if the principal directions correspond to those of the parallel and meridian through a point on the map, that graticule intersection will be a right angle.

2.10.4.2 Area scale

From the geometry of the ellipse it is also possible to derive two other measures which indicate the amount of deformation at any point on a map projection. The **Area Scale** expresses the amount of exaggeration in the size of the ellipse of distortion compared with that of the small circle on the surface of the globe. Like the particular scales, the area scale is represented by means of a decimal. If p = ab = 1.0, the ellipse of distortion is the same size as the circle on the sphere. If p = 2.0, an area in that part of the map is shown twice as large as the corresponding area on the sphere.

*Maling, D. H., 1973. *Co-ordinate Systems and Map Projections*, George Philip and Son, London, 225 pp.

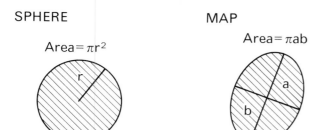

SPHERE $\quad$ MAP

Area=πr^2 $\qquad$ Area=πab

Fig. 2.59 If $\pi r^2 = \pi ab$ for all points, the map is EQUAL-AREA and the area scale = 1.0 throughout

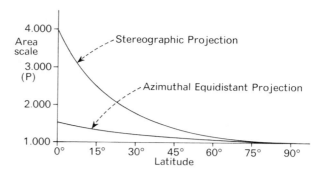

Fig. 2.61 Graphical comparison of the area scales of two azimuthal projections

2.10.4.3 Maximum angular deformation

As the name indicates, this is a measure of the extent to which the map can be used to measure angles reliably at a point. Maximum angular deformation is expressed in degrees. Thus a point for which $\omega = 0°$ is one at which any angle can be measured correctly.

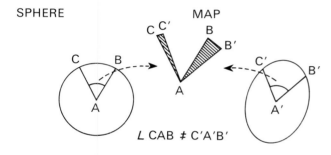

SPHERE $\qquad$ MAP

$\angle$ CAB $\neq$ C'A'B'

Fig. 2.60 ω = CAC'+BAB' is maximum angular deformation. If CAB = C'A'B' for all angles at all points on the map, the projection is CONFORMAL or ORTHOMORPHIC (and the ellipse of distortion is a circle)

2.10.5 Methods of illustrating deformation

The particular scales of a map projection can be computed and the area scale, or maximum angular deformation, can be computed from the particular scales. This is normally done for a series of graticule intersections, as in the following table, where deformation only varies with latitude.

If a map projection is a continuous representation of the spherical surface (i.e. assuming there are no edges or interruptions between a pair of points), the variation in the particular scales, area scale and maximum angular deformation may be assumed to increase regularly and continuously from one point to the other. Because of this regular variation, the deformation may be shown graphically in various ways:

— For the simpler map projections in which deformation only varies with one variable (latitude in the example on the azimuthal equidistant projection), any of the tabulated variables may be plotted as a graph.
— For the more complicated map projections, in which the deformation varies with both latitude and longitude, it is more convenient to plot isograms on the map which show equal value of area scale or equal amounts of maximum angular deformation.
— It is possible to illustrate deformation by constructing a small ellipse at every graticule intersection. This shows variation in the shape and size of the ellipses and also indicates the principal directions at each point. Consequently, a visual appraisal of the kind of deformation on the map can be made, but in

Table 2.6

Azimuthal equidistant projection particular scales, area scale and maximum angular deformation for intervals of 15° of latitude

Latitude	Particular scale along meridian	Particular scale along parallel	Area scale	Maximum angular deformation
φ	h = b	k = a	p	ω
0°	1.000	1.571	1.571	25°39'
15°	1.000	1.355	1.355	17°21'
30°	1.000	1.209	1.209	10°52'
45°	1.000	1.111	1.111	6°0'
60°	1.000	1.047	1.047	2°38'
75°	1.000	1.011	1.011	0°38'
90°	1.000	1.000	1.000	0°

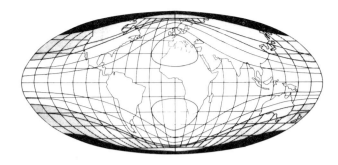

Fig. 2.62 Mollweide's projection (see 2.14.4, example 2) showing isograms for equal values of ω at 10°, 20°, 30°, 40°, 50° and > 60°

many respects it is less satisfactory than the use of isograms.

All of these techniques are useful to the cartographic editor who is required to choose a suitable projection for a map in a new atlas. In general, and notwithstanding the specialised use of certain projections for particular purposes, the best choice is the projection which shows a particular country or continent with the smallest amount of deformation. Thus by comparing similar graphs for several related projections or by comparing the patterns of the distortion isograms for several related projections, or comparing the patterns of the distortion isograms for several possible projections, we can make the choice about which one is the best to use.*

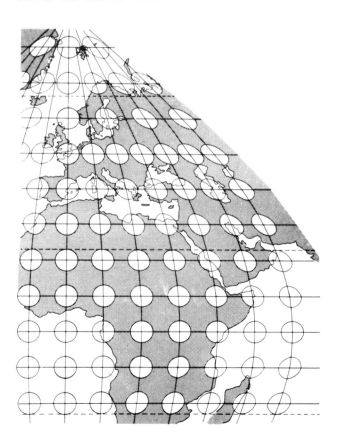

Fig. 2.63

Figure 2.63 shows part of the sinusoidal projection showing diagrammatic representation of the ellipse of distortion at the 10° graticule intersections. Note that:
- The ellipses along the Greenwich Meridian and the Equator are circles of identical size, indicating that there is no deformation along these lines. In other words, they are lines of zero distortion.
- All the ellipses on the map have the same area.
- Therefore, it is an equal-area projection (2.13.2).
- There is increasing ellipticity of the ellipses towards the north eastern part of the maps and this occurs where the parallels and meridians make oblique intersections.
- The axes of the ellipses in the north-eastern part of the map do not coincide with the directions of the parallels and meridians.

2.11 The fundamental properties of a map projection

This term refers to the following characteristics of a map projection:
- The nature of the lines of zero distortion or the number of points of zero distortion.
- The shape of the outline of the projection for the whole world or of one of the hemispheres.
- The characteristic pattern of the distortion isograms with respect to the points or lines of zero distortion and the boundary of the world or hemispheric map.

2.11.1 The cylindrical projections

If a single line of zero distortion coincides with the mapped representation of a great circle (such as the Equator or a meridian together with its antimeridian), this line will be represented at the correct scale on the map. It is geometrically equivalent to regarding the plane surface of the map as having been rolled into a cylindrical shape round the globe so that the great circle is the only part of the spherical surface which is tangential with the cylinder.

On the resulting map, which is the cylinder unrolled, the line of zero distortion appears as a straight line which is the same length as the great circle on the globe.

The world map is rectangular in outline. The particular scales increase outwards from the line of zero distortion, and therefore the isograms are also straight lines which are parallel with it.

*Maling, D. H., 1973. *Co-ordinate Systems and Map Projections*. Philips, London, 255 pp.

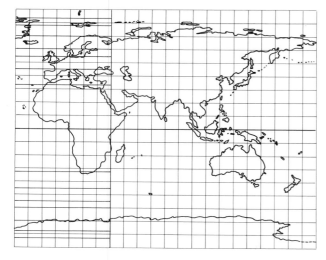

Fig. 2.64 Cylindrical equidistant (plate carrée) projection with single line of zero distortion. Isograms are for maximum angular deformation of 10°, 20°, 30°, 40°, 60°, 80° and 100°

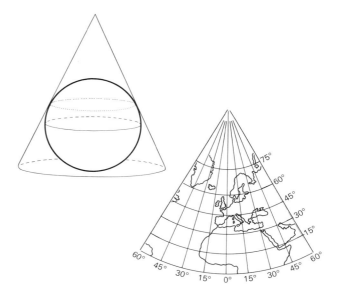

Fig. 2.66 Conformal conical projection with one standard parallel

Having thus described the fundamental properties of the tangential cylindrical projections, it is also possible to imagine the case of the secant cylinder, having two lines of zero distortion which are equidistant from the centre of the map. The shape of the world map is still rectangular and the isograms are still parallel to the lines of zero distortion, but, since these correspond to two small circles on the globe, the overall width of the projection is less than that for the tangential cylindrical projection of the same scale.

2.11.2 The conical projections

If a single line of zero distortion coincides with the representation of a small circle (such as a parallel of latitude), this line will be shown at the correct scale on the map. It is geometrically equivalent to regarding the plane surface of the map as having been rolled in the shape of a cone which is tangential to the spherical surface along the circumference of the small circle. The resulting map projection shows the line of zero distortion by a circular arc, and therefore the distortion isograms are a series of concentric circular arcs. Having thus defined a tangential conical projection, it is also possible to imagine the case of the secant cone in which there are two lines of zero distortion, each representing a small circle having a different radius on the globe. The shape of the map and the pattern of the distortion isograms are similar to those for the tangential conical projections, but there is some redistribution of the particular scales within the map.

2.11.3 The azimuthal projections

If the principal scale is preserved at only one point on the map, this is geometrically equivalent to imagining that the plane of the map is tangential to the sphere at that point. Because this is a point of zero distortion, there is no angular deformation here. Therefore, all **bearings** or **azimuths** are correctly shown at this point, hence the name azimuthal. The world or hemispheric outline of the projection is circular. The particular scales increase radially outwards from the point of zero distortion and consequently the distortion isograms are represented by the circumferences of concentric circles.

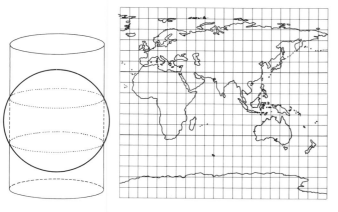

Fig. 2.65 Cylindrical equidistant projection with two lines of zero distortion (in latitudes 30° North and South). The isograms are for maximum angular deformation, with the same values as the preceding figure

53

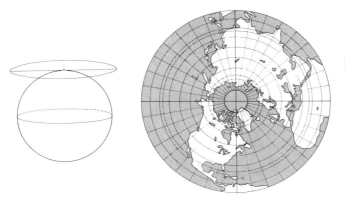

Fig. 2.67 Azimuthal equal-area projection (see 2.14.3, example 2)

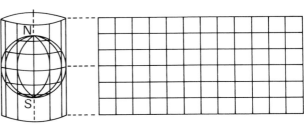

Fig. 2.69 Normal aspect cylindrical equidistant or plate carrée projection

2.11.4 Other projections

The fundamental properties giving rise to cylindrical, conical and azimuthal projections are the three simplest which we encounter in the study of map projections. There are many other kinds of representation which have two or more points of zero distortion, or two intersecting lines of zero distortion. These kinds of complication mean that it is no longer possible to imagine a simple geometrical model of the map fitted to the globe. There is, moreover, great variation in the appearance of the world map and in the patterns created by the distortion isograms. Some of these projections are briefly described or illustrated later.

2.12.1 The normal aspect

This describes the simplest form of representation, where the line of zero distortion coincides with the graticule. For example:

—**The Normal Aspect Cylindrical Projections** have the line of zero distortion coincident with the Equator. Consequently, all the parallels and meridians form a network of parallel straight lines.

—**The Normal Aspect Conical Projections** have the line of zero distortion coincident with a parallel of latitude. Consequently, all the parallels are represented by concentric circular arcs and the meridians are convergent straight lines (see Figs 2.2, 2.4, 2.66 and 2.82 for examples).

—**The Normal Aspect Azimuthal Projections** have the point of zero distortion at a geographical pole. The parallels are represented by concentric circles and the meridians are convergent straight lines.

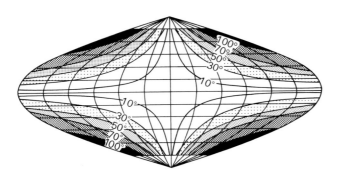

Fig. 2.68 Sinusoidal projection (see 2.14.4, example 1) showing distortion isograms for maximum angular deformation at 10°, 30°, 50°, 70° and 100°

2.12 The aspects of a map projection

This term refers to the appearance of the parallels and meridians on a map projection which, in turn, depends upon the choice of the positions of the lines or points of zero distortion with respect to the Earth. Three main possibilities are described below.

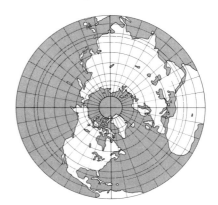

Fig. 2.70 Normal aspect azimuthal equal-area projection

2.12.2 The transverse aspect

In the transverse aspect the line of zero distortion is perpendicular to that of the normal aspect, or the point of zero distortion has been moved through 90°. For example:

- **The Transverse Aspect Cylindrical Projections** are those having a line of zero distortion which is coincident with a meridian (and its anti-meridian). Note that this gives an entirely different pattern of parallels and meridians, although the world outline and the pattern of distortion isograms is that of the normal aspect of the same projection rotated through 90°.

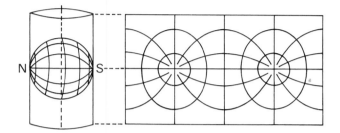

Fig. 2.71 Transverse Mercator projection

- **The Transverse Aspect Azimuthal Projections** are those in which the point of zero distortion is located on the Equator. Again, the simple systems of circles and straight lines comprising the normal aspect graticule have been replaced by more complicated curves, but the circular outline and the concentric circular isograms are unchanged.

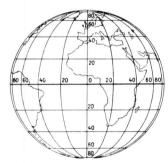

Fig. 2.72 Transverse aspect orthographic projection (see 2.14.3, example 4)

2.12.3 The oblique aspects

In the oblique aspect the line of zero distortion is any great circle or small circle which intersects the graticule. Alternatively, the point of zero distortion is located somewhere between the Equator and the poles. In each case the patterns of parallels and meridians are quite different from those for the normal or transverse aspects. However, the outline of the world map and the pattern of distortion isograms remain unaltered. Any map projection can be used in any aspect.

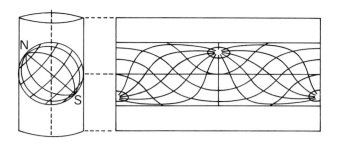

Fig. 2.73 Oblique aspect cylindrical equal-area projection

By choosing a suitable aspect for a new map, it is possible to arrange for the deformation inherent in any given projection to have the least effect within the confines of the country or continent to be mapped. The ability to change aspect is therefore an exceptionally powerful tool in the hands of the cartographic editor, who is required to make such decisions.

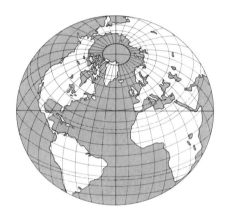

Fig. 2.74 Oblique aspect azimuthal equal-area projection

2.13 The special properties of a map projection

Although the particular scales vary from place to place on a map projection, certain mathematical relationships can be maintained between them throughout the map. This gives rise to the **Special Property** of the map projection which has an important influence upon the use of the map for certain purposes. The three most important special properties are: **Conformality**, **Equivalence**, **Equidistance**.

2.13.1 Conformality

A conformal map projection is one which satisfies the condition that the maximum and minimum particular scales are equal to one another at every point. It follows that deformation increases regularly in all directions, or that a small circle on the spherical surface is also represented by a circle on the map. Consequently, there is no angular deformation and this means that a conformal map may be used in those applications where it is necessary to measure angles from the map. The condition that a = b everywhere upon the map also means that the shapes of small areas are correctly represented, hence the use of the alternative term **Orthomorphism** to describe this property. However, it is important to realise that it is the absence of angular deformation which is the most valuable characteristic of a conformal map. This is an essential requirement for navigation charts and it is also extremely important for military use of topographical maps. It follows that all navigation charts and most topographical maps are based on a conformal projection.

2.13.2 Equivalence

This is the condition that at every point the minimum particular scale is the reciprocal of the maximum particular scale, i.e. b = 1/a, and since the area scale, p = ab, in an **Equal Area Projection**, p = 1.0 everywhere.

An equal-area map is useful for showing statistical data. Since an important element in the interpretation of statistical maps is the **density** of a variable (such as human population depicted by dots), the number of symbols per unit area creates a visual impression of density. On an equal-area map this visual impression has some meaning. On a map which does not satisfy this special property, an entirely false interpretation of density may be obtained.

2.13.3 Equidistance

This is the special property that the principal scale is preserved in the direction perpendicular to the line of zero distortion, or radially outwards from a point of zero distortion. In the example of the azimuthal equidistant projection (2.10.5), the particular scale along the meridian, h, is equal to 1.0 everywhere. The name arises from the fact that in the normal aspect used for cylindrical, conical and azimuthal projections the principal scale is preserved along the meridians and therefore all parallels on the map are equidistantly spaced. This property is not, in itself, particularly useful. A few maps are constructed on the oblique aspect of the azimuthal equidistant projection, the point of zero distortion being an important city such as London, Washington or Moscow. Such maps provide a useful method of measuring the bearings and distances to other places in the world and give a realistic picture of airline networks. However, the real advantage of the property of equidistance is that such maps have comparatively small amounts of angular deformation and the area scale does not become excessively large. The graph in 2.10.5 illustrates

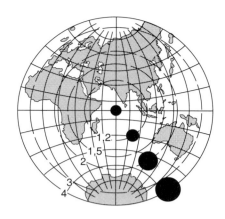

Fig. 2.75 The conformal azimuthal or stereographic projection in the transverse aspect. Note that the ellipse of distortion is always circular but that the area scale (shown by the isograms) increases radially outwards from the centre of the map

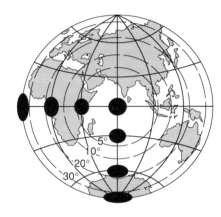

Fig. 2.76 The azimuthal equal-area projection in the transverse aspect. Note that the ellipses of distortion are all of similar area but that maximum angular deformation (shown by the isograms) increases radially from the centre of the map

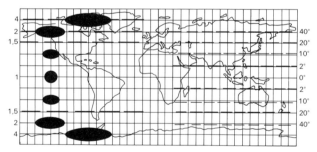

Fig. 2.77 Equidistant cylindrical or plate carrée projection. The principal scale is preserved along the meridians. Hence the minor axis of each ellipse is the same as the diameter of the circle on the Equator. Isograms on the left indicate area scale and those on the right show maximum angular deformation

this. In other words, an equidistant map is a good compromise for use when neither conformality nor equivalence are obligatory. Consequently, equidistant maps are often used in atlases as the base for general reference maps of countries or continents.

2.14 The main classes of map projection

There are seven named classes of map projections: cylindrical, conical, azimuthal, pseudocylindrical, pseudoconical, pseudoaimuthal, polyconic. Of these, the pseudoaimuthal class is unimportant and is not described further.

The following assumptions and conventions are employed in the descriptions which follow:

—The Earth is regarded as a sphere of unit radius (R = 1). Hence to compute the master grid co-ordinates from the projection co-ordinates, it is necessary to multiply by the constant factor R' which, as described (in 2.8.3.2), is the radius of a spherical globe having the same principal scale as the required map.

—The following algebraic notation is used: φ = latitude; (x,y) = Cartesian projection co-ordinates (x is the ordinate); λ = longitude; (r, θ) = polar projection co-ordinates; and χ = co-latitude = $90° - \varphi$.

Where the angles φ, λ or χ occur in an equation and are not expressed as some trigonometric function (such as sin φ), it is to be assumed that the angle is measured in radians.

2.14.1 Cylindrical projections

Fundamental properties:
1. Line of zero distortion is a great circle on the sphere represented by a straight line on the map.
2. Particular scales increase in the direction perpendicular to the line of zero distortion. Therefore, isograms are straight lines parallel to the line of zero distortion.
3. Outline of the world map is rectangular.

Normal aspect graticule:
4. The Equator is the line of zero distortion.
5. Parallels are represented by parallel straight lines all of which are the same length as the Equator.
6. Geographical poles are lines having the same length as the Equator.

7. Meridians are equidistant parallel straight lines which are perpendicular to the Equator.
8. All the graticule intersections are perpendicular and correspond to the principal directions.
9. Maximum and minimum particular scales therefore coincide with those along the meridians and parallels.
10. The only difference between cylindrical projections possessing different special properties is in the spacing between the parallels.

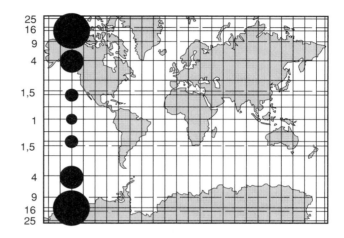

Fig. 2.79 The conformal cylindrical or Mercator's projection. Isograms are for area scale

Examples:

Cylindrical equal-area projection: x = sin φ, y = λ (see 2.8.3.3 for illustration).

Cylindrical equidistant (plate carrée) projection: x = φ, y = λ (see 2.11.1 for illustration).

Mercator's projection (conformal cylindrical):
$$x = \log_e \tan\left(\frac{\pi}{4} + \frac{\varphi}{2}\right), \quad y = \lambda.$$

14.01c

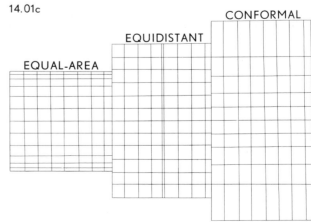

14.01a

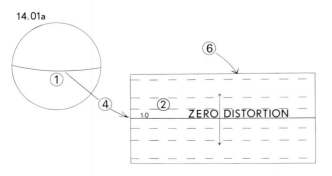

Fig. 2.78

Fig. 2.80

Modifications:

The commonest modification to the cylindrical projections is to employ two lines of zero distortion which, in the normal aspect, are two parallels which are equidistant from the Equator. A number of different cylindrical projections are distinguished by name on this basis alone (see Fig. 2.65 for illustration).

2.14.2 Conical projections

Fundamental properties:

1. Line of zero distortion is a small circle which is represented by a circular arc on the map.
2. Particular scales increase in the direction perpendicular to the line of zero distortion. Therefore, the distortion isograms are circular arcs concentric with the line of zero distortion.
3. Outline of the world or hemisphere is fan-shaped.

Normal aspect graticule:

4. Line of zero distortion is a parallel of latitude, known as the **Standard Parallel** (denoted φ_0 in the equations).
5. Other parallels are concentric circular arcs.
6. The geographical pole is either a point or a short circular arc.

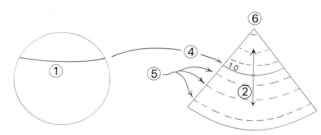

Fig. 2.81

7. The meridians are straight lines which converge at an angle which is less than the difference in longitude between them.
8. All the graticule intersections are perpendicular and correspond to the principal directions.
9. Maximum and minimum particular scales therefore coincide with those along the meridians and parallels.
10. Differences between the individual projections depend, in part, upon the spacing of the parallels and, in part, upon the angle of convergence of the meridians.

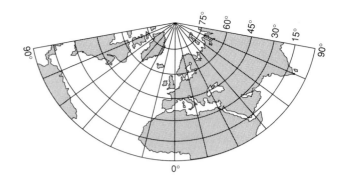

Fig. 2.82 Conical equal-area projection with standard parallel 50°N

Examples:

Equidistant conical projection: $r = \cot \varphi_0 + (\varphi_0 - \varphi)$, $\theta = n\lambda$, where $n = \sin \varphi_0$ and is known as the **Constant of the Cone** (see 2.2.2 for illustration).

Equal-area conical projection: $r = \dfrac{2}{\sqrt{n}} \sin \dfrac{\chi}{2}$,

$\theta = n\lambda$, where $n = \cos^2 \dfrac{\chi_0}{2}$.

Conformal conical projection:

$r = \tan \chi_0 \left[\dfrac{\tan \frac{1}{2} \chi}{\tan \frac{1}{2} \chi_0} \right]^n$,

$\theta = n\lambda$,
where $n = \cos \chi_0$

(see 2.11.2 for illustration).

Modifications:

The commonest modification is the use of two standard parallels rather than one. This has the effect of redistributing the particular scales on a map, thereby reducing the deformation towards the edges.

58

2.14.3 Azimuthal projections

Fundamental properties:
1. One point of zero distortion which is the origin of the projection co-ordinates.
2. Particular scales increase radially outwards from this point. Therefore, the distortion isograms are concentric circles.
3. Outline of the world or hemispheric map is circular.

Normal aspect graticule:
4. The point of zero distortion is the geographical pole.
5. Parallels of latitude are concentric circles.
6. The meridians are straight lines converging at the pole and the angles between the meridians on the map correspond to the difference in longitude between them on the sphere.

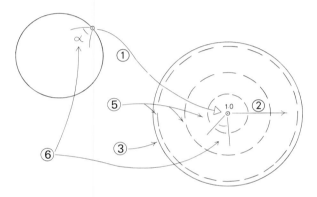

Fig. 2.83

7. All the graticule intersections are perpendicular and correspond to the principal directions.
8. Maximum and minimum particular scales therefore coincide with those along the meridians and parallels.
9. Differences between the individual projections depend solely upon the spacing of the parallels.

Examples:
1. Azimuthal equidistant projection: $r = \chi$, $\theta = \lambda$.
2. Azimuthal equal-area projection: $r = 2 \sin \dfrac{\chi}{2}$, $\theta = \lambda$ (see 2.11.3, 2.12.3, 2.13.2 for illustration).
3. Stereographic projection (conformal): $r = 2 \tan \dfrac{\chi}{2}$, $\theta = \lambda$ (see 2.13.1 for illustration).
4. Orthographic projection: $r = \sin \chi$, $\theta = \lambda$ (see Fig. 2.72 for illustration).
5. Gnomonic projection: $r = \tan \chi$, $\theta = \lambda$.

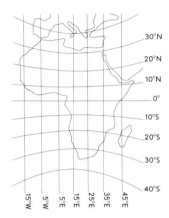

Fig. 2.85 Gnomonic projection (transverse aspect). This projection has the special property that all great circle arcs on the sphere are shown by straight lines on the map

Modifications:
As a logical development of the concept of tangent and secant cylinders or cones (2.11.1–2.11.2), it is also possible to modify the tangent plane of the azimuthal projections into a secant plane. This gives rise to a standard circle, which is a line of zero distortion in some latitude other than the pole. This is used sometimes with the stereographic projection, but it is much less common than the corresponding modifications to the cylindrical and conical projections.

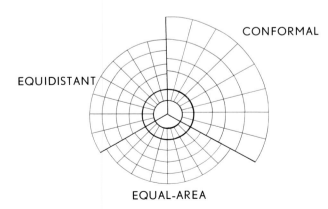

Fig. 2.84

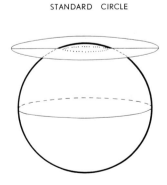

Fig. 2.86

2.14.4 Pseudocylindrical projections

Fundamental properties:

1. Either two lines of zero distortion corresponding to two perpendicular great circle arcs or two points of zero distortion.
2. Particular scales increase outwards from the lines or points of zero distortion. Frequently the distortion isograms have fairly complicated patterns.
3. World outline may be elliptical, rectilinear (diamond-shaped) or formed from two contiguous parabolae, hyperbolae or sine curves (sinusoids).

Normal aspect graticule:

4. The lines of zero distortion are often the Equator and the central meridian of the map. Versions with two points of zero distortion only have these located on the central meridian.
5. Parallels of latitude are parallel straight lines.
6. The meridians are curves similar to those defining the world outline. For example, a world map enclosed within an ellipse has elliptical meridians, and one formed from two sine curves has meridians which are also sine curves.
7. Graticule intersections along the Equator and the central meridian are usually perpendicular, but elsewhere they are not. Consequently, the principal directions do not usually correspond with the graticule.

Examples:

1. Sinusoidal projection (equal-area): $x = \varphi$, $y = \lambda \cos \varphi$.
2. Mollweide's projection (equal-area): $x = \sqrt{2} \sin \psi$, $y = \frac{\sqrt{8}}{\pi} \lambda \cos \psi$ (ψ is the auxiliary angle to be found from the equation $\sin 2\psi + 2\psi = \pi \sin \varphi$*).
3. Parabolic projection (equal-area): $x = \sqrt{3} \sin \frac{\varphi}{3}$, $y = \lambda \left[\frac{3}{\pi}\right]^{1/2} \left[1 - \frac{4x^2}{3\pi}\right]$.

*For methods of solving this equation see Wagner, K. H., 1949. *Kartographische Netzenwürfe.* Bibliographisches Institut, Leipzig, 262 pp.

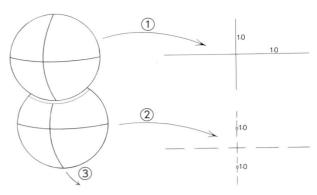

Fig. 2.87

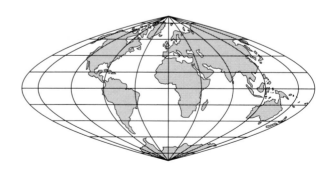

Fig. 2.88 Three equal-area pseudocylindrical projections. Top: sinusoidal projection; centre: Mollweide's projection; bottom: parabolic projection

60

Modification:

The commonest modification to pseudocylindrical projections is to replace the geographical pole with a line which is some simple ratio of the length of the Equator. Usually this is one-half, but other ratios have been described. The effect of introducing a pole line is to reduce angular deformation towards the edges of the world map. Another common modification is the **recentred** or **interrupted** version of the projection.

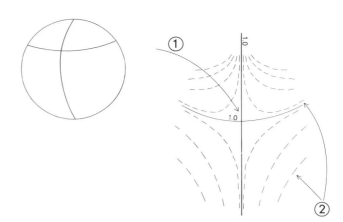

Fig. 2.91

3. World outline is roughly bell-shaped and in some examples it is heart-shaped.

Normal aspect graticule:

4. The lines of zero distortion are a standard parallel and the central meridian.
5. Parallels of latitude are concentric circular arcs.
6. The geographical pole is a point.
7. Meridians are curved symmetrically about a rectilinear central meridian.
8. Graticule intersections are perpendicular along the lines of zero distortion but not elsewhere.

Example:

Bonne's projection (equal-area):

$$r = (\cot \varphi_0 + \varphi_0) - \varphi, \ \theta = \frac{\cos \varphi}{r}.$$

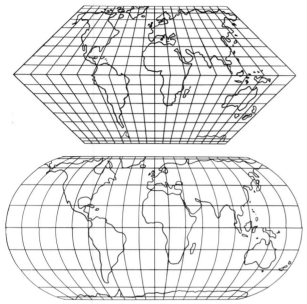

Fig. 2.89 Two equal-area pseudocylindrical projections with pole line. Top: Eckert II projection (rectilinear meridians); bottom: Eckert IV projection (elliptical meridians)

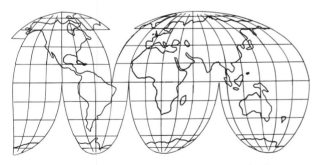

Fig. 2.90 Interrupted or recentred Mollweide projection

2.14.5 Pseudoconical projections

Fundamental properties:

1. One line of zero distortion, which is a small circle on the globe, intersecting a second line of zero distortion which is a great circle on the globe.
2. Particular scales increase outwards from the two lines of zero distortion so that the distortion isograms are curves located symmetrically about these lines.

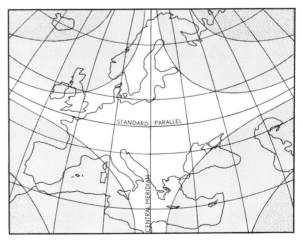

Fig.2.92 Bonne's projection with standard parallel 50°N for a map of Europe. Isograms for 1° and 5° of maximum angular deformation.

Modifications:

None, apart from the choice of the standard parallel. The limiting case, where $\varphi_0 = 90°$, is the heart-shaped Stab-Werner projection.

2.14.6 Polyconic projections

Mathematically speaking, the polyconic class represents the general case of all map projections, of which the six classes already described are special cases.

Fundamental properties:

1. One or more lines of zero distortion or one or more points of zero distortion.
2. Particular scales increase outwards from the points or lines of zero distortion so that the distortion isograms are usually curves.
3. World outline may be almost any kind of regular geometrical figure.

Normal aspect graticule:

4. Usually the Equator and the central meridian are represented by straight lines. The other parallels and meridians are generally curved. The examples of rectilinear or circular parallels and meridians are characteristic of the other classes which have been described.
5. Graticule intersections may be perpendicular throughout the map, but usually they vary from place to place.

Examples:

1. The polyconic projection: $x = \varphi + 2 \cot \varphi \sin^2 \left[\dfrac{\lambda \sin \varphi}{2} \right]$, $y = \cot \varphi \sin (\lambda \sin \varphi)$ (see 2.10.1 for illustration).

2. Hammer-Aitoff projection (equal-area):

$$x = \sqrt{2} \, \frac{\sin \varphi}{1 + \cos \varphi \cos \frac{1}{2}\lambda},$$

$$y = 2\sqrt{2} \, \frac{\cos \varphi \sin \frac{1}{2}\lambda}{1 + \cos \varphi \cos \frac{1}{2}\lambda}$$

(see 2.8.3.3 for illustration).

Fig. 2.93 Aitoff's projection, which exemplifies a polyconic projection bounded by an ellipse

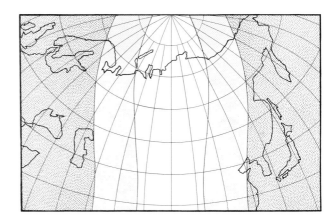

Fig. 2.94 Simple polyconic projection used for a map of northern Eurasia. Distortion isograms are for $\omega = 1°$ and $5°$

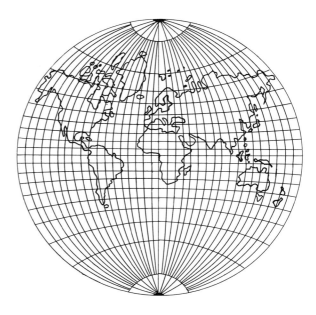

Fig. 2.95 Van der Grinten's projection, which exemplifies a polyconic map of the world enclosed within a circle

Useful modern references on mathematical cartography

English language

Maling, D. H., 1989. *Measurements from Maps*. Oxford: Pergamon Press.†

Maling, D. H., 1991. *Coordinate Systems and Map Projections*. Oxford: Pergamon Press.†

Richardus, P. & Adler, R. K., 1972. *Map Projections for Geodesists, Cartographers and Geographers*. Amsterdam: North-Holland.‡

Snyder, J. P., 1987. *Map Projections, A Working Manual*. USGS.†

Snyder, J. P. & Voxland, P., 1989. *An Album of Map Projections*. USGS.†

Steers, J. A., 1962. *An Introduction to the Study of Map Projections*, 13th ed. London: University of London Press.*

French language

Marchant, R., 1961. *Notions sur la Théorie des Projections Cartographiques à l'Usage des Agents Cartographes*. Bruxelles: Institut Géographique Militaire.*

Reignier, F., 1957. *Les Systèmes de Projection et Leurs Applications*. Paris: Institut Géographique National.‡

Reyt, A., 1961. *Leçons sur les Projections des Cartes Géographiques*. Paris: Institut Geographique National.‡

German language

Fiala, F., 1957. *Mathematische Kartographie*. Berlin: VEB Verlag Technik.‡

Heissler, V. & Hake, G., 1970. *Kartographie I*, 4th ed. Berlin: Sammlung Goschen Band 30/30a/30b, de Gruyter.*

Wagner, K.-H., 1962. *Kartographische Netzentwürfe*, 2nd ed. Mannheim: Bibliographisches Institut.†

Russian language

Garaevskaya, G. A., 1955. *Kartografiya*. Moscow: Geodezizdat.*

Ginzburg, G. A. & Salmanova, T. D., 1964. *Posobie po Mathematicheskoy Kartografii*. Moscow: Trudy TsNIIGAiK, Vyp 160. Izd. Nedra.†

Graur, A. V., 1956. *Mathematicheskaya Kartografiya*. Leningrad: Idz. Leningradskogo Universiteta.†

Meshcheryakov, G. A., 1968. *Teoreticheskie Osnovy Mathematicheskoy Kartografii*. Moscow: Izd. Nedra.‡

Solov'ev, M. A., 1969. *Mathematicheskaya Kartografiya*. Moscow: Idz. Nedra.†

Urmaev, N. A., 1962. *Osnovy Mathematicheskoy Kartografii*. Moscow: Trudy TsNIIGAiK, Vyp 144. Geodezizdat.‡

* Elementary standard.
† Intermediate standard.
‡ Advanced standard.

Chapter 3

THEORY OF CARTOGRAPHIC EXPRESSION AND DESIGN

B. Rouleau

CONTENTS

3.1 The aims of cartographic expression and map reading

3.1.1 What purpose does a map serve? Cartographic information

Spatial information can be communicated to a reader by means of visual devices such as maps, diagrams and graphs, the presentation and appearance of which are controlled by a variety of graphic principles. These illustrative methods are used to depict interrelationships and differences between elements or sets of data and a plane surface, i.e. one exhibiting the orthogonal dimensions of length and breadth. However, the main aim of cartography is to represent the correct spatial location of data on this plane surface. A map can therefore be said to be a graphic depiction, on a two-dimensional/plane surface (normally a piece of paper), of the location of geographical phenomena and their relative positions within a given space.

Information and language. As is the case with other graphic constructions (diagrams and graphs), maps employ a form of visual language to communicate items of information. This differs from spoken language (articulation), written words (literacy), or musical expression which are not necessarily always immediately, completely, or universally understood, and where differing interpretations may be possible. The cognition of visual language is, conversely, an instantaneous, complete and universal process provided that the rules of its grammar are understood (see 3.1.2).

Instantaneousness: relationships among data are represented by means of three variables. Two of these, x and y, define the positions of points on the plane surface. The third relates to the visual appearance of these points. Interrelationships between these variables can be perceived instantaneously and totally.

When looking at the symbols included in Fig. 3.1 it is immediately possible to distinguish two different groups of dots—one large, the other smaller: visual language allows an overall impression to be formed at once.

Completeness: a map must contain all of the elements of information which it is intended to communicate. These items must be homogeneous and carefully and accurately positioned.

Universality: cartographic language obeys universal rules relating to visual perception; anybody viewing Fig. 3.1 will notice that one group of dots is of a significantly larger size than the other.

Consequently, the cartographic image can be said to have only one meaning, whereas a phrase or sound can have several, as is exemplified by the possible translation of a word or sentence into different languages.

Why use a map? A map is both useful and efficient:
- —if it provides a graphic answer to a question posed, i.e. if it communicates the necessary information;
- —if it is made taking into account the general laws of visual perception.

A map provides answers to questions. It demonstrates the locations of phenomena and communicates relationships existing between data in terms of their differences, orders or proportions (see 3.2.3). The method of cartographic presentation selected must provide answers to the questions posed.

What sorts of questions can be asked?

- —at a general level: how are industries distributed throughout a region?
- —at a more specialised level: where are textile industries located?
- —at a specific, local level: what industries have been established at a particular place (Fig. 3.4)?

3.1.2 Visual perception and map reading

A map will communicate information, more or less immediately, relating to a number of levels of perceived detail depending on:
- —the degree of complexity of the employed symbolisation;
- —the correct application of the rules of visual perception.

Maps requiring study. In Fig. 3.4 all of the industrial information tabulated in Fig. 3.5 has been located and symbolised, but the map only provides a general impression. In order to discover specialised detail with respect to the location of chemical plants, it is necessary to study all of the included symbols sequentially. Figure 3.2, illustrating land values in eastern France, demonstrates a similar situation.

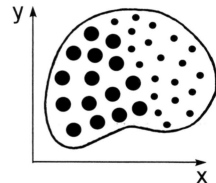

Fig. 3.1

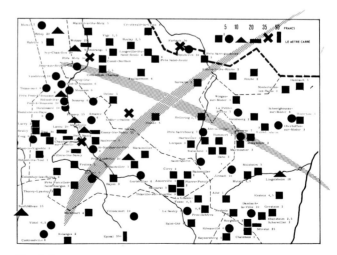

Fig. 3.2

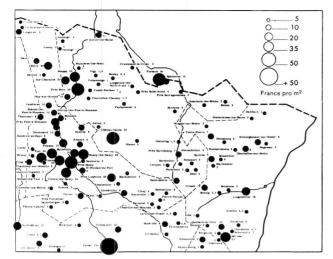

Fig. 3.3

Textile	Mecanical	Hosiery	industries localities
●			ARCHES
	●		AUDINCOURT
			BACCARAT
			BARR
			BRIEY
			CHARMES
		●	CHATEL
		●	DARNEY
			DIEUZE
●	●		EPINAL
●	●		LUNEVILLE
●	●		NANCY
●	●		OBERNAI
	●		PONT-A-MOUSSON
●			RIBAUVILLE
●	●	●	ST DIE

Fig. 3.5 Some of the information on manufacturing represented in Fig. 3.6

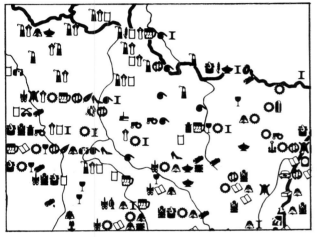

Fig. 3.4

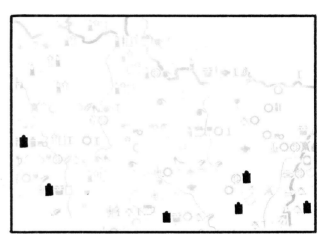

Fig. 3.6

Maps providing immediate answers. Figure 3.6 exemplifies only factories involved in the textile industry, and thus the reader is immediately able to see where these are situated. From Fig. 3.3 it is possible to see at once where the most valuable land is located, and the reader can also go on to discover values at particular places. The differences between Figs 3.2 and 3.3 result, primarily, from the use of poor symbolisation on the former. Devices normally employed to differentiate between various classes of phenomena have been used in an attempt to show variations in proportions relating to one topic. In addition, rules governing visual perception have been sensibly applied in the design of proportional symbols for Fig. 3.3.

3.1.3 What sort of map should be made? Types of maps

Is it essential to illustrate all of the available data, or could this be simplified to facilitate the communication process? A complex map may incorporate representation of all information concerning 'n' factors (Figs 3.8 or 3.9). A simplified map results from the processing of data relating to the indicators (Fig. 3.10).

Inventory maps? The superimposition or juxtapositioning of 'n' characteristics on one map permits its user to read only detail relating to a specific point. Answering a question concerning data distributed over the whole map is virtually impossible, or at best will necessitate close study for a considerable period of time.

This type of map is complex and allows the answering of only certain types of questions — Where is this city? How can I get from one place to another? In these situations the character of the land surface plays an important role (topographic or road maps . . .) (Fig. 3.7).

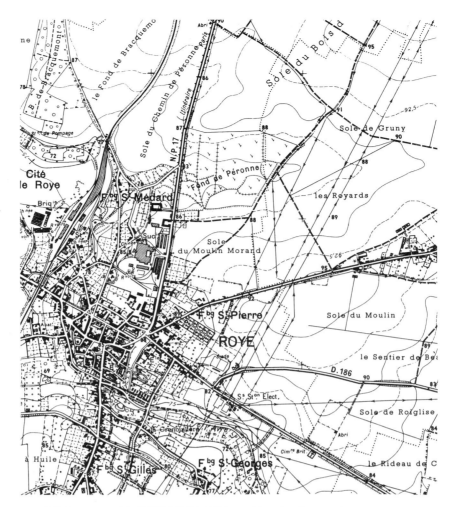

Fig. 3.7 A topographic INVENTORY MAP requiring detailed study

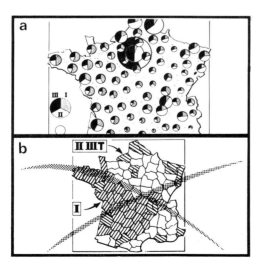

Fig. 3.8(a)
A complex THEMATIC INVENTORY MAP
requiring detailed study. (b) A useless map!

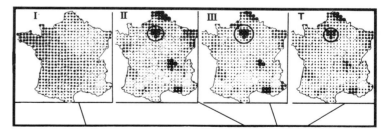

Fig. 3.9 ANALYTICAL MAPS: immediate answers are provided by a series of maps each displaying a single characteristic

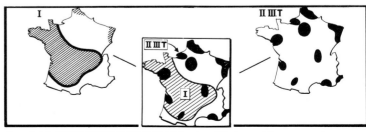

Fig. 3.10 COMMUNICATION MAPS: simplified graphic summaries of detail

Analytical maps? These are detailed maps depicting only one geographical attribute (Fig. 3.9). The representation of 'n' characteristics will require the generation of a series of maps at the same scale and with similar base elements. Each can be read at the specific, local level, but also provide an answer at the general level. They are, above all, suited to detailed analysis and the regional comparison of phenomena represented. The simplification of detail when presented in this way makes the maps valuable tools for research purposes.

Communication maps? In this type of presentation, original information, obtained from the comparison of 'n' analytical maps (Fig. 3.10), is summarised into one or a limited number of general categories, and displayed as a composite product. It answers questions at all levels.

Before starting compilation it is necessary to decide on the type of map required. Data have to be handled and processed in different ways depending on the intended purpose of the end-product.

3.2 Data capture, documentation and processing

3.2.1 Sources of information

Whether involved in the generation of topographic or thematic materials a cartographer will, neces-sarily, be involved in the processing of spatial information prior to its intended graphic display. As a result of this operation it should be possible to position details accurately with respect to their actual geographical location, and also to demonstrate the essential characteristics of data and their relationship with one another. By utilising a variety of techniques, the cartographer attempts to provide a 'graphic representation of information which is appropriate for communication to a user'. The precise positioning of spatial detail is the responsibility of a surveyor. However, a knowledge and awareness of the characteristics of each of the items of included information can be obtained from a variety of documentary sources. The cartographer should be capable, in specific instances, of 'researching', 'analysing' and processing such data prior to their graphic portrayal.

Amongst the most consistently employed sources of information for both large- and medium-scale mapping are those resulting from the recording of 'observations made in the field'. These relate particularly to the production of topographic, geological, lithological and land-use maps. Other relevant information may be obtained from the interpretation of 'aerial photographs', or terrestrial imagery captured by remote sensing apparatus carried in orbiting satellites, e.g. for the production of vegetation maps.

Alternatively, new mapping can be derived from 'existing materials' at large scales, or compiled from information published in books or journals.

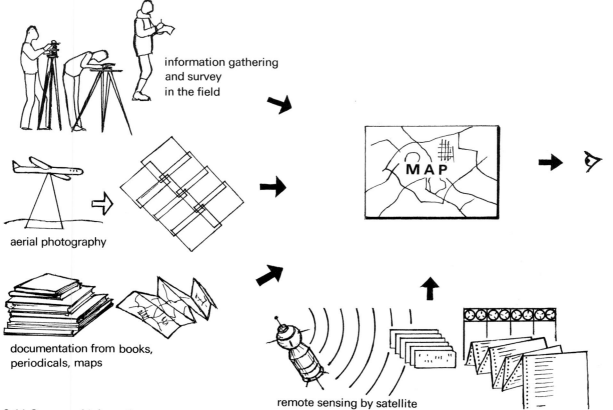

information gathering and survey in the field

aerial photography

documentation from books, periodicals, maps

remote sensing by satellite

Fig. 3.11 Sources of information

Finally, an ever-increasing amount of 'quantitative/statistical detail' is becoming readily available, on an annual or even daily basis, and provides the most important or sole source of data for use in the compilation of thematic mapping depicting topics such as climate, hydrology, demography, agricultural and industrial production, etc. (see Fig. 3.11).

In a growing number of countries, particular agencies are responsible for the compilation and storage of these types of details in the form of databases. Records are maintained on magnetic tapes or discs, and much of the information is made readily available for use by cartographers and other interested parties.

It is always necessary for those involved in map preparation to evaluate the accuracy and relevance of available statistics prior to their processing and, ultimately, graphic portrayal in map form.

3.2.2 Data

To enable the creation of a cartographic image it is essential that the compiler has access to spatial data, i.e. information relating to specific, known locations.

The spatial location of data may be merely 'geographic' (see Fig. 3.12) or 'administrative' (see Fig. 3.13). In this second case the information refers to defined administrative areas which also serve as statistical units when mapping topics such as population density or change. A third type of spatially located detail is directly attributed to the specific point or place at which the data were collected, for example a meteorological station (see Fig. 3.14).

Further, data occurring in space can be 'continuous' and evenly distributed, or 'discontinuous' and scattered randomly. They can also be collected in a 'quantitative' form as numerical values (absolute quantities, proportions, percentages, etc.), or as data which cannot be measured and are termed 'qualitative' (land use or settlement types, tourist infrastructure, etc.).

All of this information can also consist of different components or 'characteristics', each of which constitutes a variation in the data. For example, the different characteristics present within a land use unit or population group.

A 'variable' is an item of data which is not constant and can be subject to a number of changes in structure or function, or exhibit a variety of characteristics. The magnitude of the characteristics is termed the 'range' of the variable. In the case of quantitative variables the related characteristics may be comprised of 'continuous values' (quantities) such as temperature or water run-off, or 'discontinuous values' exemplified by factors like the number of inhabitants, statistics concerning administrative areas, and amounts of

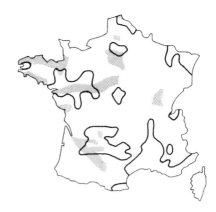

Fig. 3.12 Distribution of principal French vegetable and fruit producing regions; representation of geographical location, i.e. the actual extent of areas

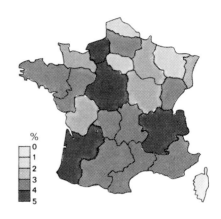

Fig. 3.13 The statistical data collected often relates to administrative units, the boundaries of which hinder geographical analysis of the phenomena. This map illustrates the annual variation of urban population in France, by regions, between 1962 and 1968

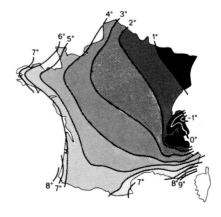

Fig. 3.14 Average January temperatures: data are collected at individual meteorological stations and isotherms, at equal interval steps, are interpolated from them. They serve to demonstrate a continuous increase, as also happens when contours are used to represent relief

rainfall.

In all instances 'a map is only of use when it exemplifies differences existing between the location and characteristics of at least two phenomena, or a minimum of two characteristics of a single phenomenon'.

3.2.3 Relationships between data

A map is intended to communicate information, but the desired 'message' will only be evident if the relationships between included data elements or sets are immediately perceivable. These can occur at three different levels:

- quantity relationship: this allows data to be measured, and provides the answer to the question 'How many?' (e.g. inhabitants, tonnes, areas, etc.) (see Fig. 3.15);
- relative order: data can give answers to questions such as 'Which is the first, second, third . . . etc.?' These data can be numerical, as is illustrated in Fig. 3.16, where information has been classified, by amount, into groups with consecutive values. Alternatively, data may be non-measurable and relate, for example, to chronological order (periods of time) or hierarchical succession (administrative ranking) (see Fig. 3.17):
- differences: in some instances data can be neither measured nor ordered, but are only 'nominally' different one from another (types of crops, industries, etc.) (see Fig. 3.18).

In certain cases differences make a 'selection' possible. For example, the distribution of data by families, groups or categories which are not specifically ordered (groups of industrial activities which are related, or vegetational species) (see Fig. 3.19).

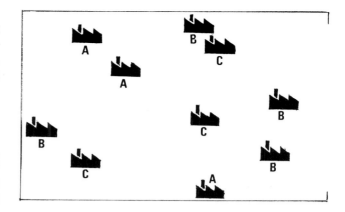

Fig. 3.16 Numbers of employees by classes: A = 5–20; B = 21–50; C = 51–100 (typical 'classified' information)

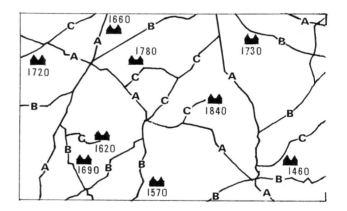

Fig. 3.17 Example of hierarchical and chronological ordering: A = major road; B = secondary road; C = farm track (typical 'ranked' information)

Fig. 3.15 Numbers of inhabitants: typical 'quantitative' data

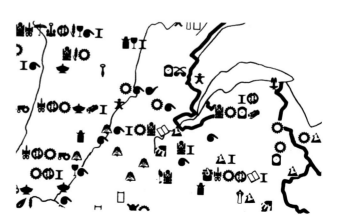

Fig. 3.18 All illustrated details appear to be at the same visual level and are only 'nominally' different

71

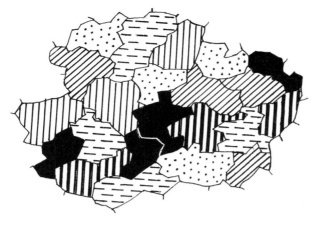

Fig. 3.19 Areas represented in the same way immediately appear to be related ('selection')

3.2.4 Data processing

Maps can be compiled to illustrate a 'visual synthesis' resulting from the processing of original data. This can be achieved, for example, by using a visual classification system involving ordinal matrices (distinction between large and small, major or minor); an image card-index of the type developed by J. Bertin (Figs 3.20(a) and 3.20(b)); or by the employment of a technique requiring the construction of triangular graphs, which serve to illustrate the number of dominant factors necessary for representation, and whether these are ordered series or random characteristics (see Fig. 3.21).

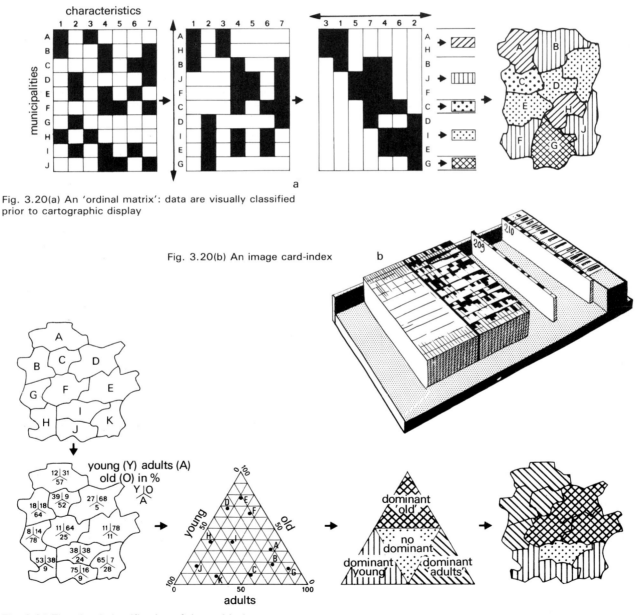

Fig. 3.20(a) An 'ordinal matrix': data are visually classified prior to cartographic display

Fig. 3.20(b) An image card-index

Fig. 3.21 The visual classification of data with three characteristics by means of a 'triangular graph', and their cartographic representation

The informational content of any cartographic illustration must be compiled taking into account the quantity of appropriate detail which can reasonably be expected to be communicated to a user at the intended production scale. If too large a scale is selected, and too little information included, the attention of the potential reader may become dissipated. Conversely, if the chosen scale is too small, and the content too detailed, a reader may well experience difficulty in distinguishing and interpreting the various categories of information. In consequence, the intended communication process will again become ineffective.

3.3 Graphic representation

3.3.1 Information transmission: the cartographic language*

Information communicated by a map must be easy to see, understand/interpret and remember. In order to achieve these aims the cartographer must employ an appropriate language, which is:
- 'visual': it follows the general rules relating to visual perception and is immediately observable;
- 'universal': it must be readily understood by an intended reader;
- 'graphic': the map is a tangible object which can be stored and compared with others contained in a collection or atlas. It can be reduced, enlarged, reproduced or transmitted.

3.3.2 Positioning

Before selecting a method of graphic representation the spatial location of available data must first be analysed. 'Detail positioning is the recording, on the plane surface of the map, of the geographical locations of data' (see Fig. 3.22). The flat map is a representation of a part of space, and all of the known locations of positions within this can be projected onto the plane in terms of their 'x' and 'y' co-ordinates. In addition, it is also possible to locate phenomena, defined in numerical terms, which are not immediately visible on the Earth's surface.

Each point can be considered as laying at the intersection of two co-ordinate axes which may be either geographical or geometrical/Cartesian. It can be digitised, coded and stored as part of a digital database.

*The structure of graphic language was defined by J. Bertin in *Semiology of Graphics*, Madison, Wisconsin, 1983.

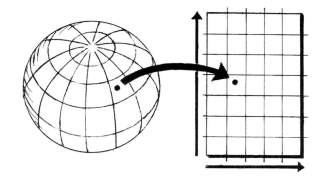

Fig. 3.22 'Positioning': the spatial location of data can be accurately plotted and represented on a plane surface

Positioning systems. The location of features can be referred to in terms of 'points, lines or areas':
- point positions are defined as being related to specific locations in space;
- linear positioning is explained in terms of the connection of two specific points;
- area location may be accomplished by creating a bounding line interconnecting a minimum of three points which do not lie in a straight line. This boundary delimits the physical extent of a phenomenon in space (see Fig. 3.23).

Characteristics of positioning. The specific location of a phenomenon is an element of the data with which a cartographer is provided. He may not change it, it is therefore 'fixed'. However, he may need to modify the positioning of the data (i.e. its location and graphic portrayal on the plane surface of the map), depending on the scale employed, by means of 'generalisation'. For instance, a feature represented as a cluster of dots at a large scale may, necessarily, be shown as an area symbol at a medium scale or a single point symbol at a very small scale (see Fig. 3.24). Generalisation relating to positioning is essential in order that a map remains clear and legible. The technique is frequently employed in topographic cartography and in the compilation of base detail for use in thematic mapping.

3.3.3 The bases of the language—the graphic element

Simple graphic devices representing point, line and area information form the core of the cartographic language. These drawn elements have no specific meaning when considered in isolation, but this situation changes when they are used as representative symbols.

3.3.4 Symbol design

A graphic 'symbol' is a device selected by the cartographer in order to represent data, a phenomenon or a concept. Consequently, it is the

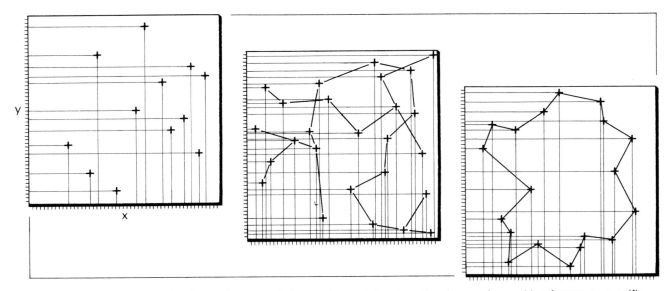

Fig. 3.23 The positions of point, line and area symbols are always defined, on the plane surface, with reference to specific points which can be precisely located in terms of 'x' and 'y'. These positions can be numerically coded and classified. Subsequently, they are plotted or stored as part of a database. Automated mapping/Geographic Information Systems (GIS) are founded on this principle

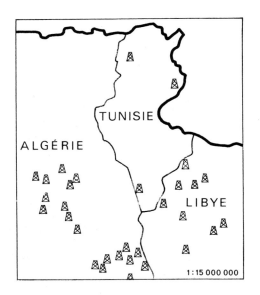

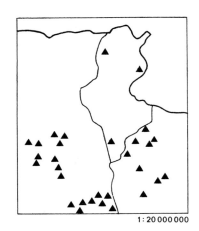

Fig. 3.24 Symbol modification. Although not changing a concept, or the related data, it may be necessary to alter the appearance of a symbol as a result of decrease in scale

most fundamental element of cartographic language and design. Map symbols must be constructed to demonstrate clearly:

- —the precise geographical location of the features they represent;
- —the relationships existing among them with respect to quantities, ranking, similarities and differences.

Symbols incorporated in a map are particularly differentiated one from another by virtue of their individual positions.

A symbol can be of the 'point-variety': constructed to demonstrate a particular location or the centre of a distribution (more specifically they are devices whose generalised shape does not precisely duplicate that of the actual feature on the ground); 'linear': comprised of an axis joining two points; 'areal': drawn with reference to a boundary line corresponding to the physical extent of a phenomenon on the Earth's surface (see Fig. 3.25).

'Symbol design is independent of position': a point symbol may be employed to denote a specific location; as a constituent element of a line; or as part of an area pattern. The same is true of linear devices (see Fig. 3.26).

'Symbol design is also independent of the area to which it relates': a point symbol may occupy much more space on the map than the actual, physical extent of the item which it represents

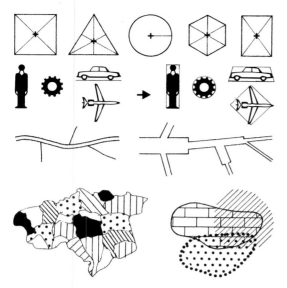

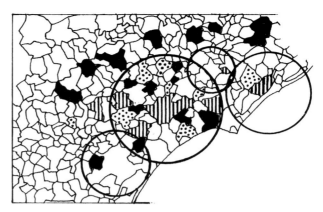

Fig. 3.27 Point representation and area classification

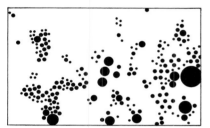

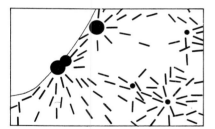

Fig. 3.25 Symbol variety: geometric point symbols can be modified in terms of their character and size; pictorial symbols can be enclosed by geometric boundaries; line symbols; area symbols. All of these types can be employed to demonstrate either qualitative or quantitative detail

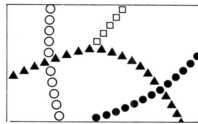

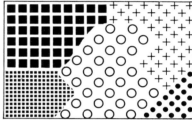

Fig. 3.26(a) Point symbols can be used to demonstrate a specific location; to form a linear symbol; or constitute elements of an area pattern

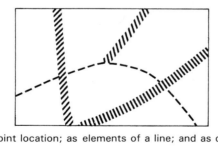

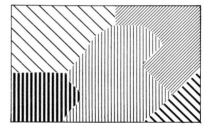

Fig. 3.26(b) Linear symbols relating to point location; as elements of a line; and as components of an area pattern

(e.g. a circle proportional to the population of an important city with a small geographical area). Conversely, an area symbol may be very small (e.g. detail relating to the area of a small municipality incorporated on a map at a scale of 1 : 2 000 000 (Fig. 3.27)).

'Generalisation of symbols': it was earlier mentioned that, when the scale of a map is enlarged or reduced, it may sometimes be appropriate to generalise and alter the positions of symbols. This change may also make necessary the modification of the character of the employed graphic devices which, nevertheless, continue to represent the same data (see Fig. 3.24).

3.3.5 Visual variables

These are methods relating to the modification of the visual appearance of graphic elements or symbols used on a map. The purpose of this is to demonstrate differences between, or relationships among, various classes of data (Fig. 3.28). The characteristics of included symbols which can be manipulated are their size, texture, value, colour, orientation and shape.

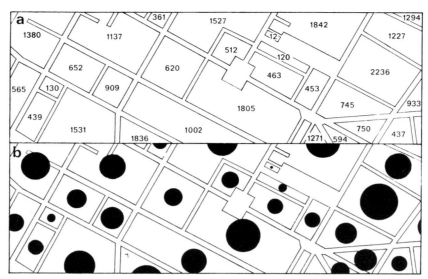

Fig. 3.28 Numerical values shown by figures: (a) There is no apparent visual difference. (b) The same values are shown by proportional symbols. Differences can be perceived immediately

3.3.5.1 Size

The size variable can be expressed in terms of either the length or area of a symbol. Sometimes it is also possible to suggest differences with respect to volume, although the eye can only perceive the 'x' and 'y' dimensions on a plane surface: variations in volume are normally difficult to compare one with another.

As a rule, possibilities for changes in size are virtually unlimited, and in fact the only constraint is man's ability to perceive certain minimum thresholds of linear or areal size differences (see also 3.4.4).

3.3.5.2 Texture and structure

Within a symbol, simple graphic elements can be regularly distributed to create a specific 'texture'. This term refers to the shape of the elements, and 'structure' describes their ordered spatial arrangement (see Fig. 3.29).

By employing variations in the texture and/or structure of the elements comprising different

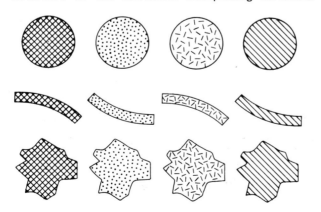

Fig. 3.29 The generation of different textures and structures within point, line and area symbols in order to create visual differences

devices, it is possible to demonstrate a clear distinction between either individual or groups of symbols.

3.3.5.3 Value

The term value variable applies to the differences in light intensity which are perceived by the eye as shades of grey ranging between white and black (see Fig. 3.30). In practice, especially when these variations are reproduced during map printing, changes of texture are employed. By modifying the size, and consequently the spacing of the graphic elements (dots or lines) comprising a symbol, a difference in visual value can be created. Value is thus a relationship between the amount of a surface which is covered by the graphic elements and the white space between them. This is expressed as a percentage, with a clear white surface corresponding to 0% and a black one 100%. In the latter case, the elements completely cover an area which is then described as being 'saturated'. Film screens at 10% intervals are commercially available (i.e. at 10%, 20%, 30%, 40%, 50%, 60%, 70%, 80%, 90%). The human eye experiences considerable difficulty in attempting to differentiate values of less than 10% between two adjacent areas on a map (see Fig. 3.31). When there is an equilibrium between white and black, the value is rated as 50%.

'Structure and vibration effect': when the graphic elements and white background exhibit an equivalent value (50%), a vibration or flickering effect occurs. This is particularly noticeable when the dots or lines comprising a structure are big enough to be individually distinguished. The eye tends to flicker between the black and white areas of apparently similar importance. Textures of this density create an unpleasant sensation and their

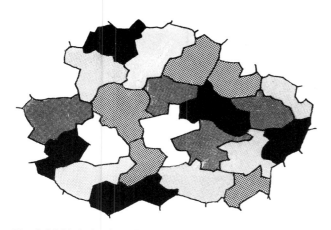

Fig. 3.30 Variation in lightness or perceived value

use should be avoided. They tend to give the impression that the features to which they are applied are more important than is actually the case (alternating bands, concentric circles, thick pecked lines, etc.) (Fig. 3.32).

3.3.5.4 Grain

When the dimensions of the graphic elements forming a symbol are modified by photographic enlargement or reduction, a variation in grain results. This causes 'no change in structure or value', with the relationship between the marks and the white background remaining the same. Grain is a 'size variation' of composing elements, or of point or line symbols distributed over a surface or area (Fig. 3.33).

10% 20% 30% 40% 50% 60% 70% 80%

20% 50% 80%

Fig. 3.31 Relationships between white and black expressed in terms of percentages

Fig. 3.32 Visual vibration or 'flickering' resulting from graphic elements occupying 50% of the available white background

Fig. 3.33 Variations in grain: employed graphic elements (lines or dots) are ordered with respect to apparent size. However, their relationship with the percentage of the white background covered remains constant

77

3.3.5.5 Colour

Variations in hue (red, yellow, blue, orange, green, purple, etc.), which the eye can discern when looking at elements comprising a multi-coloured map, are not produced by using straightforward graphic techniques or employing special features relating to the design of symbols, but rather by changing the printing inks used during the reproduction process.

What is colour? It can be explained as a 'physical sensation' which is produced in the eye by rays of decomposed light. The colour radiations, which correspond to the printing inks used on the map, are 'components of white light which is emitted by the sun'. This consists of 'photons' each of which has a specific spectral wavelength.

Solar light contains radiations at all wavelengths of the spectrum, and appears white when it contains a mixture of all of its components. However, in certain circumstances the visible wavelengths can be separated. This occurs when a beam of light is broken up by using a prism or filters (in nature we would term this a rainbow effect). It then becomes evident that the various constituents are arranged according to a well-defined sequence: violet-blue, blue, green, yellow, red-orange, red. The human eye cannot perceive any light at wavelengths of less than that for violet-blue, or greater than that of red. Between these limits (approximately 400 to 700 nanometres or thousandths of a micron) coloured rays follow one another sequentially according to defined wavelengths.

The latter correspond to a portion of the solar spectrum where each colour exists in a pure (monochromatic) state and at its maximum intensity. A physical sensation may be visually experienced, either as a result of a white body reflecting coloured light, or by a coloured body reflecting white light. All of the colours that the eye can perceive relate to large or small portions of the spectrum, the wavelengths of which are characterised by the 'hue' of the colour.

Fundamental colours — primary colours. The three 'fundamental colours' are violet-blue, green and red-orange, and these are positioned at approximately equal distances, one from another, within the visible portion of solar spectrum wavelengths. They are complementary to the 'primary colours': 'cyan' (blue), yellow and 'magenta' (purple-red) (see Fig. 3.34).

Combination of colours. It is also possible to combine colours in terms of different value percentages (tints). By using the three primary colours, and by superimposing screens, it is feasible to generate a wide variety of colours in all sorts of tints. Colour charts demonstrate that shades and tints of orange can be produced by combining diverse tints of yellow and magenta; greens can result from the mixing of cyan and yellow; and shades and tints of violet are obtained by the superimposition, during the printing process, of different tints of cyan and magenta. The amalgamation of solid (100%) versions of the three primary colours will produce black; and the use of all three in an unsaturated form results in the creation of a brownish tint.

Additive and subtractive syntheses. If circles consisting of red, green and blue light are projected onto a white screen, the superimposition of red and green produces yellow; the overlapping of green and blue creates cyan; and a combination of blue and red generates magenta. When all three of these fundamental colours (red, green and blue) are overlaid, they give white. This is an additive synthesis, and is illustrated in Fig. 3.35(a).

If coloured filters are introduced in front of a

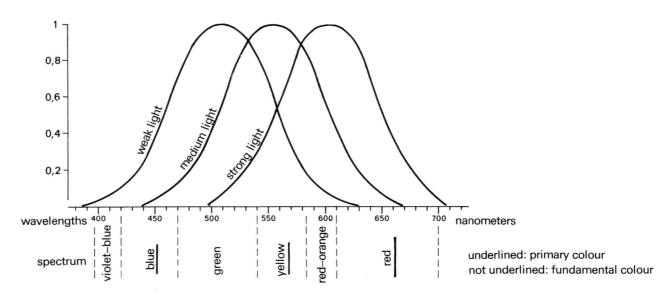

Fig. 3.34 The perceivable spectral colour sequence, and the influence of light source intensity upon it

beam of white light, the superpositioning of magenta and yellow will give red; magenta and cyan will produce blue; and cyan and yellow create green. The combined use of all three filters results in the generation of black. This is described as the subtractive synthesis (see Fig. 3.35(b)):

1. The 'additive primaries' (red, blue and green) are of particular importance when designing maps for projection or use with colour television.
2. The 'subtractive primaries' (magenta, cyan and yellow), plus black, conform with the colours of the basic printing inks used in the production of multi-colour maps.

Colour, value and saturation. All colours can be made, subject to variations in their lightness or value (i.e. made to appear lighter or darker), when they are displayed in combination with more or less light as a result of using screens. 'Pure' or 'solid' colours (without any white) are termed 'saturated'. It is impossible for the majority of map readers to distinguish more than three variations of yellow saturation; four or five of green or orange; six of red; and up to eight for blue and violet.

Perception of colour brightness. The apparent colour of an object or map element will be perceived by the eye in terms of the 'amount' and 'composition' of the light shining on it. If the ray is complete (i.e. if it incorporates the three fundamental colour wavelengths), it will not alter the original colour of the object. If, however, light is incomplete, the apparent colour will be modified. For instance, yellow and red detail will appear grey when viewed under orange illumination (subway lights). Colours used must be selected with care when a map is being produced, or is to be employed under artificial lighting conditions (see Fig. 3.36)!

The 'intensity of light' will also influence colour appearance and perception. Colours with long wavelengths (yellow to red) will appear brighter when viewed under strong light, and softer in subdued illumination (at dusk). The opposite is true for blue and violet which have short wavelengths (see Fig. 3.34).

Psychological aspects of colour. The emotions and sensations experienced by a map reader when studying coloured documents have always provided, consciously or unconsciously, a guide for

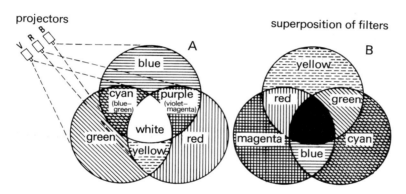

Fig. 3.35 The principles of colour combination. (a) Superimposition of coloured lights projected onto a white screen: the 'additive synthesis'. (b) Transparent coloured filters superimposed and positioned in front of a beam of white
light projected onto a white screen: the 'subtractive synthesis'

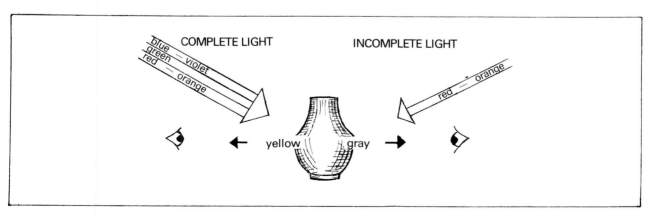

Fig. 3.36 Colour modification under artificial lighting conditions

cartographers. Connotative colours are those which remind a reader of features of the natural environment (green for forests and vegetation in general; blue for precipitation and factors relating to hydrology or the sea; yellow and orange for cereal crops or elements pertaining to drought, sand, etc.). Further distinctions are made between warm colours (red, yellow) and those which are cool (blue, violet) or fresh (green).

The so-called warm colours are used to represent map details which the cartographer intends should appear more important or nearer to the eye, while those considered cool depict less important facts and elements which are further from the eye. These general 'conventions' are, in fact, based on the map user's actual perception of the employed colours. Because of the differences in their wavelengths colours reach the eye at different rates: thus some are seen 'earlier' and as being 'nearer' (red and colours with long wavelengths), whilst others are seen 'later' or as 'further away' (blue and other colours with short wavelengths).

Conventional practice in selecting colours (which may vary in different parts of the world) should normally be respected in order to assist the local understanding of map materials.

Colour interaction. Individual areas of colour will appear more or less pure or intensive depending on the environment within which they are positioned. A colour appears clearer and more vivid when it is surrounded by white, and darker or duller when encircled by black (see Fig. 3.39). The representation of narrow linear elements (area boundaries, contours, roads) in yellow should be avoided, particularly when these details have necessarily to be superimposed on darker colours.

3.3.5.6 Orientation

It is possible to make a distinction between symbols by means of introducing differences in orientation. However, this variable can only be applied to textures or linear symbols, and thus its usefulness is limited.

Orientation can be used as either a graphic or a geometric variable, and employed to represent different classes of information (i.e. human factors or vegetation types) (see Fig. 3.37). Additionally, it may serve to demonstrate a precise geographical direction as an element of a flowmap (see subsection 3.5.3).

Orientation is the only one of the graphic variables which can provide an effective representation of dynamic phenomena such as direction, movement, immigration, emigration, etc. (see Fig. 3.38).

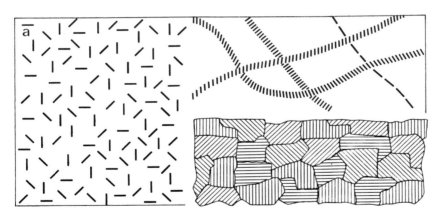

Fig. 3.37 The employment of variations in linear element orientation (vertical, horizontal and at 45°) to promote visual differentiation between point, line and area symbols

Fig. 3.38 Representation of movement (rural migration) by using differently oriented linear symbols. (a) Radial orientation. (b) Concentric orientation

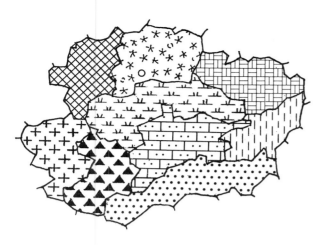

Fig. 3.39 Area patterns

3.3.5.7 *Shape*

Variations in shape or form—not to be confused with the boundaries of specific geographical locations—consist of changes in the outline of a symbol. These modifications apply only to point symbols or, in certain circumstances, lines. They never relate to the outline of an area symbol which represents a defined location and so cannot be altered (see Fig. 3.40).

The only possibility available for the expression of differences of shape within an area lies in the introduction of textures, point or line symbols which are distributed at regular intervals across the area. The shapes of these must remain consistent within a particular boundary, but may vary from area to area with each form representing a defined category of data. Variations in symbol patterns are commonly employed in mapping topological surfaces (see Fig. 3.39). Shapes used in cartography are termed pictorial or representative when they suggest, in a general way, the actual character of the information illustrated. However, much of the data mapped cannot be

seen in real life and so will not be shown by this method. Geometric shapes such as circles, squares and triangles are becoming increasingly commonly used for the depiction of detail, especially statistical data.

3.4 Rules of cartographic language

3.4.1 Cartographic information and the application of visual variables

The information contained in a map is only communicated effectively when the relationships among included data elements or sets are demonstrated clearly and simply. These associations are emphasised by an appropriate use of visual variables. Each of the latter 'allows only a certain level of relationship among the data elements or sets to be represented'.

3.4.1.1 *Representation of quantity relationships*
Variations in symbol 'size' provide the only means of representing quantity differences, as only these changes 'can be measured'. We refer to continuous variation when the dimensions of each symbol are made proportional to the amount it represents. A map compiled using this technique will contain as many differently sized symbols as there are quantities to be distinguished (see Fig. 3.28).

The variation becomes discontinuous when the amounts to be illustrated are grouped into classes, each of which is subsequently depicted by a symbol with a size proportional to it. The dimensions of the various classes of symbols should, in all cases, be proportional to precise quantities. In Fig. 3.40 these relate to the average values of each of the groupings.

None of the other visual variables can ever be used to demonstrate quantities or the relationships among them.

Fig. 3.40 Variations in circle size demonstrating the value of classes relating to open field cultivation (*Atlas de Paris et de la Région Parisienne*, 1967)

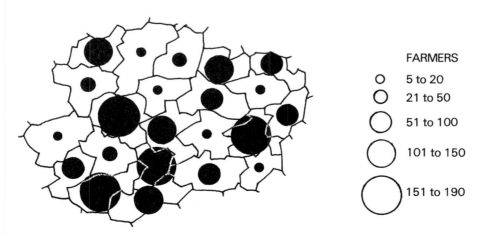

FARMERS

○ 5 to 20
○ 21 to 50
○ 51 to 100
○ 101 to 150
○ 151 to 190

3.4.1.2 Representation of order relationships

The employment of a variety of tonal 'values' on a map 'always' suggests that hierarchical order relationship exists among included data, and provides an immediate visual impression (see Fig. 3.41). Similarly, variations in 'size' can be used to show order relationships among data, since it is always possible to classify changes in symbol dimensions (see Fig. 3.42). Contrasts in grain (see paragraph 3.3.5.4), which are essentially size differences, can also be introduced to represent an order relationship. However, when they are employed as an area symbol (an area consisting of a specific texture) these variations do not seem to suggest value differentiation, and therefore will not represent order relationships (see Fig. 3.43).

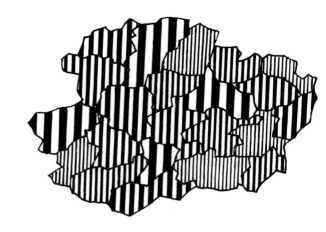

Fig. 3.43 Quantitative classes depicted by variations in the grain of linear elements: If only the 'width' of lines is considered, it is easy to 'rank' areas, in terms of apparent visual importance, with respect to the evident dominance of employed grain. If only the total area occupied by line symbols is considered (i.e. as area symbols), all of the administrative regions demonstrate the same visual value. Consequently, it is impossible to define an 'order relationship'

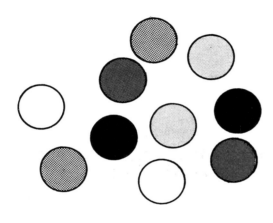

Fig. 3.41 Employing various grey tonal values can facilitate the differentiation of individual circle symbols. The darkest are perceived and mentally grouped first (selection), and subsequently others are ranked according to their apparent decreasing tonal intensity

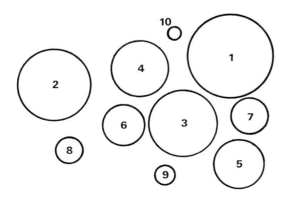

Fig. 3.42 The sizes (areas) of these circles are proportional to particular quantities. In addition, they serve to express an order relationship between data

Colours do not conform to a standard sequence, although a natural progression from one to the next is evident when they are arranged according to their respective positions in the visible spectrum: yellow–orange–red, or yellow–green–blue–violet. This is termed harmonic gradation.

The remaining visual variables are unsuitable for the illustration of order relationships.

3.4.1.3 Representation of differences

All of the visual variables express dissimilarities among data elements or sets, but some serve merely to demonstrate qualitative differences. For instance, symbols contrasted only in terms of shape or form, texture and orientation appear to occupy the 'same level' of importance on a map. It is not possible to measure them or arrange them in order. Nevertheless, it is necessary to distinguish them at two levels of perception:

—'selection level': differences in 'colour' and 'texture' are always employed to depict families or categories of related data. These groups cannot be classified unless, as was suggested above, colours are used in their correct spectral sequence, or if textures are combined with variations of value (see Fig. 3.44);

—'classification level': change of shape (which can only be related to point symbols) and orientation (solely applicable to linear or elongated devices) may only be used to depict simple differences between data. It is not relevant to any kind of quantitative or order relationship. Variations of orientation enable selectivity in only very particular cases (see Fig. 3.45).

Fig. 3.44 These circles can only be differentiated in terms of employed 'texture', which enables their grouping into five unspecified categories: 'selection level'

a   b

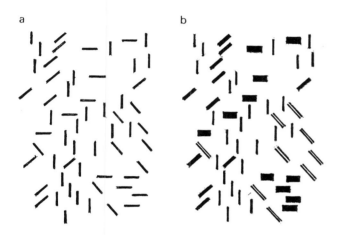

Fig. 3.45(a) The 'orientation' variable is used alone: selection is very difficult. (b) 'Orientation' is used in conjunction with 'size', and in consequence selection is easier

3.4.2 Visual variables as related to methods of symbolisation

The use of visual variables is specifically related to symbols and not to their geographic location. Application is thus linked to the particular methods of point, line and area symbolisation by which spatial information can be represented. An appropriate method of depiction is selected with reference to the nature and positioning of the phenomenon to be illustrated. The actual placement of information on the map is defined by its geographical location and the map scale.

'Variations in size' can only be applied to 'point or line symbols' exhibiting a geometric shape, as only devices of this type can be measured (radius of a circle, length of the side of a square, width of a line, etc.). In all of these cases the symbol size is made proportional to the value which is to be demonstrated.

Graphic devices of these sorts can be made to show details relating to a zone or area. For example, a proportional circle can illustrate the population of an administrative unit or, alternatively, repeated point or line symbols (forming regular patterns with different textures) can demonstrate differences in density from one area to another. However, an area, when represented by an area symbol, cannot change in size because this is fixed by the physical/geographical extent of the phenomenon (see subsection 3.3.2). Pictorial point symbols (see subsection 3.3.4) can only be varied in size when closely related to geometric representations (see Fig. 3.25).

As far as 'variations in value, texture and grain' are concerned, the methods of symbolisation utilised (point, line or area), and the related geographical locations, are of little importance. Nevertheless, the employed graphic devices must be of a sufficient size to allow users to perceive these inbuilt differences. It has already been demonstrated that 'variations in orientation' are only applicable when dealing with 'linear or elongated symbols', and that 'shape variations' are essentially restricted to 'point symbols' (although lines may be subjected to such modifications in certain limited cases).

'Colour variations' are effective when used with all three types of symbols, whatever their size, but the degree of success depends on their 'surroundings' (effect of adjacent or surrounding colours, simultaneous contrast, etc.) (see paragraph 3.3.5.5 and Fig. 3.46). All of the visual variables can be used in any location—even differences in size, orientation and shape relating to area representation. In this instance, point symbols (varying in size and shape) or lines (modified with regard to length, width, orientation and shape) are evenly distributed over the area in question.

	SYMBOL		
	P	L	Z (A)
size	X	X	
value	X	X	X
colour	X	X	X
texture	X	X	X
grain			X
orientation		X	
shape	X		

Fig. 3.46 The potential applications of visual variables relating to point, line and area symbols

Therefore, the employment of visual variables is influenced by:
- the relationships existing between data elements or sets (quantity, sequence, differences);
- the types of symbols (point, line, area) used to represent these data.

3.4.3 Combination and effectiveness of the visual variables

The perception and reading of map detail always occurs at different stages, which conforms with the various 'levels of perception'. Geographic distribution is the first thing to be seen, and this is followed by the perceiving of groups of symbols employing the same visual variables. If the map has been produced in monochrome, size or value differences will be noticed first (i.e. variations in quantities or an ordered sequence in the data being depicted). In the case of a multi-coloured document, contrasts will often appear first. In all instances groups of symbols with the same texture do not stand out as especially evident; those with the same orientation are even less readily visible; and finally, devices exhibiting the same shape are the most difficult to perceive.

Therefore, 'visual variables have different levels of effectiveness' which are universally understood and accepted. However, it is not possible to classify these levels too rigidly, as the successfulness of each visual variable also depends on the size or area of the symbol and its environment. It is, however, necessary for the cartographer to take into account the relative efficiency of the variables, particularly when the 'hierarchy of importance or interest' of the data to be displayed must be emphasised in the most effective way possible.

'Potential scope of visual variables': the possible applications of these techniques vary, one from another, in terms of their success in depicting a particular number of different data sets or elements. It is possible, at least in theory, to introduce unlimited size variations into a map (the only controlling factors being overcrowding or the minimum perceivable size differences exhibited by included symbols) (see subsection 3.4.4).

Normally it is impossible to distinguish more than 7 or 8 tonal values of grey colours or textures within a map environment; with respect to orientation, only 4 differences can be detected. Modifications of symbol shape can be unlimited, provided that pictorial devices are employed, but only 6 or 7 geometric possibilities can be accommodated. The choice of visual variables is determined with reference to the number of characteristics of a particular variation which require representation.

'Combination of visual variables': in order to enhance the possibilities of graphic effectiveness, several visual variables, for example colour and value, may be used in association (climatic, population and hypsometric tint mapping). Similarly, size and orientation can be combined (line screens differentiating between different areas), as can size and shape (the location of commercial and industrial establishments), or texture and value (area representations).

In such cases two types of relationships between data elements or sets are represented. Size and value will always be dominant, and serve to express the quantitative or sequential relationships among data; these associations, which would not have been emphasised by using one variable in isolation, now stand out clearly (see Fig. 3.47).

Fig. 3.47 The combination of visual variables: 'size + texture + value + shape'

The 'cartographic image' is thus developed and can be explained as: a graphic, symbolised assembly established with reference to accurate geographical location (determining its position) and by the type of relationships which the map author intends to illustrate.

3.4.4 Limits of visual perception: legibility rules

A cartographic document should be easily readable, under standard lighting conditions, by a user with normal eyesight. Map printing must not exercise a detrimental effect on the quality of the original image. Provided that these requirements are met, they will ensure the legibility of the smallest included details (threshold of perception); the distinction between adjacent elements of information (threshold of separation); and the discrimination of the smallest variations between the symbols included on the map (threshold of differentiation). Consequently, when the cartographer designs a graphic representation he or she must always take the following criteria into consideration.

3.4.4.1 Threshold of perception

This relates to the minimum size of a graphic element which can be viewed, with the naked eye, under normal circumstances. In theory, for an isolated element drawn on white paper, this means:
- 0.1 mm for the diameter of a point symbol;
- 0.06 mm for the width of a linear element.

However, in practice one should always take into account the density of the detail comprising a cartographic image; the use, in some instances, of a coloured background or printing surface; and also the possible employment of reduction factors.

The following standards are more normally followed:
- minimum diameter of a point symbol: 0.2 mm;
- minimum width of a line symbol: 0.1 mm (0.8 mm for certain elements in exceptional circumstances);
- minimum length of the side of a solid square: 0.4 mm;
- minimum length of the side of an open square: 0.6 mm (see Fig. 3.48).

3.4.4.2 Threshold of separation

It is essential to take into account the minimum amount of separation necessarily introduced between adjacent symbols in order to ensure their individual discrimination, with the naked eye, under normal viewing conditions. As is the case with a camera lens, the ability of the human eye is limited with respect to detail distinction. By considering test images, a viewer will quickly discover that there is a minimum perceivable distance which can be employed to separate graphic elements, and that below this threshold they will appear to merge. In quantitative terms this amount is 0.2 mm for parallel lines or densely screened areas (see Fig. 3.49).

3.4.4.3 Threshold of differentiation

There is also a minimum observable degree of distinction between symbols or devices of approximately the same size which can be discerned under standard viewing conditions. In order to provide the correct delineation, clarity, and subsequently unambiguous interpretation and appreciation of these differences, a cartographer should avoid use of:
- shapes which, potentially, are too similar (equilateral and isosceles triangles, circles, hexagons and octagons);
- tint screens of apparently equal visual density (especially with respect to the illustration of small areas);
- similarly sized symbols of a standard shape.
  It is essential to take these factors into account
- especially in the field of thematic cartography.
  The considerations quoted relating to the thresholds of perception and separation are particularly important with regard to problems relating to generalisation — especially when one is involved in topographic mapping (see Fig. 3.50).

3.5 Systems of cartographic representation

Simple devices or symbols can be combined to produce systems of graphic depiction relating to specific types of representation (point, line or area). These techniques are employed to satisfy, by virtue of their visual effectiveness, the requirements of particular types of data — especially statistical information.

3.5.1 Choosing a representational method

A variety of different methods are available for the illustration and display of data, but selection will be controlled by:
- the types of details to be shown (qualitative or quantitative) and their interrelationships;
- the geographical location of the data;
- the intended scale of the eventual map.

These considerations influence:
- the selection of relevant visual variables;
- the positioning of employed symbols.

Available systems of representation can be related to particular types of mapping, and have inherent limitations with respect to their effective display of information. For example, a point symbol may be employed to demonstrate precise location and accurate quantitative detail, but cannot be used to display the geographical extent of a phenomenon. Similarly, only a line symbol can be used to depict factors relating to spatial networks (hydrography, traffic, transportation, etc.) and quantities attributable to them. The area symbolisation of statistical data involves the grading of values, with information being grouped into numerical classes.

3.5.2 Point representation mode: locational cartography

When it is necessary for the cartographer to represent the spatial distribution of population, he or she will almost automatically think of using point symbols, as these are the most evocative in terms of demonstrating the combination of location and quantity. However, there are in fact additional possibilities for the use of point symbols to show this information. For example:
- 'A point symbol with a defined unit value' (the dot map). This involves the use of a 'single point symbol' of a standard size and value (e.g. 1 dot with a diameter of 1 mm = 10 inhabitants).

  Alternatively, it would be possible to employ a series of symbols with various sizes and standard values. For example, 1 small dot

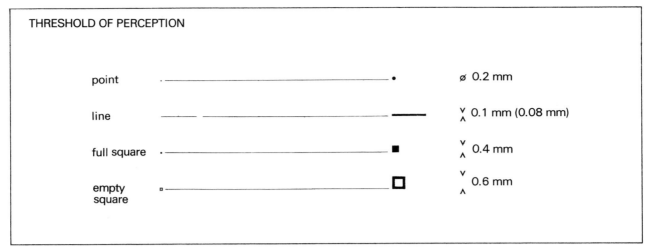

Fig. 3.48 The use of thresholds in graphic illustration

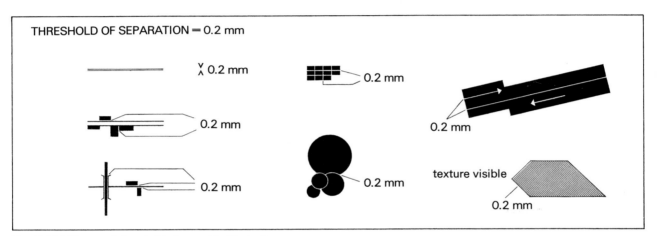

Fig. 3.49 The use of thresholds in graphic illustration

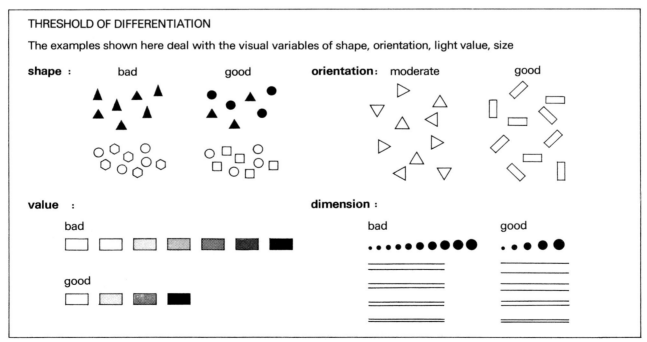

Fig. 3.50 The use of thresholds in graphic illustration

= 5; 1 medium sized dot = 50; and 1 large dot = 500 inhabitants (see Fig. 3.51). These points are accurately located and represent population centroids. This is the only method whereby the shape of agglomerations can be suggested, and also, in a general way, their spatial distributions.

If accurate positions are not known, but merely the region in which the population is located, point symbols (dots) of equal value can be evenly distributed over the area concerned. This technique creates some interesting density effects, but the resultant representation is more abstract (see Fig. 3.52).

— 'Repeated unit value symbols'. These are a variant of the previously described method.

Small circles, or preferably squares, each representing a unit (or a specific number of units) and indicating the total amount of a phenomenon present, are positioned next to one another in the centre of the appropriate area. The resultant image is abstract in nature, but if the symbols are carefully arranged in groups they can be easily compared (see Fig. 3.53).

— 'Proportional symbols'. A circle (square or rectangle), with an area proportional to the quantity being represented, is positioned in the centre of the district to which it refers. The areas of the individual symbols are related to one another in the same way as the values they represent (see Fig. 3.54).

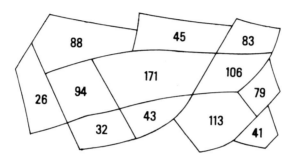

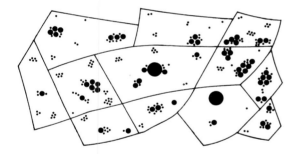

Fig. 3.51 Example of the point representation mode

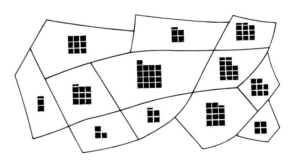

Fig. 3.52 Example of the point representation mode

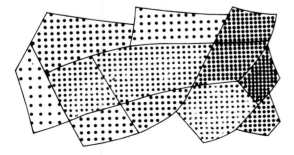

Fig. 3.53 Example of the point representation mode

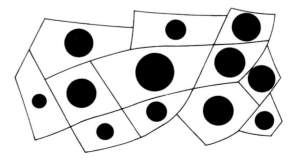

Fig. 3.54 Example of the point representation mode

When

N = the maximum number to be represented;
n = one of the other values in the series;
S = the area of the circle representing N;
s = the area of the circle representing n;
R = the radius of the circle representing N;
r = the radius of the circle representing n;

then $\dfrac{s}{S} = \dfrac{n}{N}$ or $\dfrac{\pi r^2}{\pi R^2} = \dfrac{n}{N}$,

thus $r = \dfrac{R\sqrt{n}}{\sqrt{N}}$.

In the case of squares or triangles, with sides C for N and c for n, the employed formula is

$$c = \frac{C\sqrt{n}}{\sqrt{N}}.$$

Before making any decisions the cartographer must first determine the appropriate value of R or C with respect to space available in the district containing N. In order to calculate r or c the use of a slide rule or pocket calculator is recommended. Square root tables should also be consulted. However, there is also a method which does not require the undertaking of any calculation. This involves the use of a 'nomogram', a graphic device constructed especially for this purpose (see Fig. 3.55).

—'Value classification'. The various numerical quantities relating to a data series may be grouped into classes, each of which can be represented by a standard symbol which applies to all values with similar characteristics. The employed system of representation will consist of standard sized dots proportional to the quantities incorporated in specific classes (the Bertin system illustrated in Fig. 3.56), or individual circles with areas relating directly to the mean values of the illustrated data sets (see Fig. 3.57).

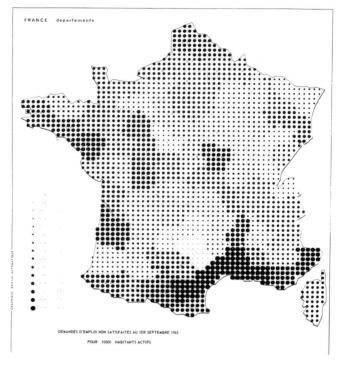

Fig. 3.56 Job applicants

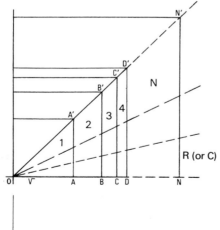
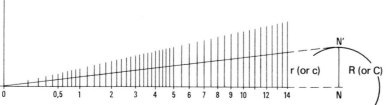

Fig. 3.55 Construction and use of a 'nomogram'. The values 1, 2, 3, 4 . . . N are represented by squares with areas of 1, 2, 3, 4 . . . N, i.e. with sides = √1(OA), √2(OB), √3(OC), √4(OD) . . . √N(ON). The triangles OAA', OBB', OCC', ODD' . . . ONN' are similar and proportional to the square root of the numbers; similarly, the lengths AA', BB', CC', DD' . . . NN' are also proportional. The NN' value is selected first and equals R or C; the line N'O is then drawn and defines all the values of 'r' (= AA', BB', CC', DD' . . .). Thus in order to construct a nomogram: (1) the values of the square roots of the numbers are plotted on an axis from an origin (O); (2) a perpendicular NN' = R (or C) is drawn at the value N; (3) the line N'O, defining the value 'r' (or 'c'), is drawn corresponding to any value measured in OX (nomogram after H.X. Lenz Cesar)

88

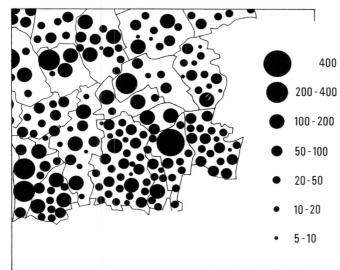

●	400
●	200 - 400
●	100 - 200
●	50 - 100
●	20 - 50
●	10 - 20
·	5 - 10

Fig. 3.57 Distribution of landowners

It is advisable to restrict the use of these techniques to mapping intended for the general illustration and rapid interpretation of classes or categories of information, as their employment results in an unavoidable loss of specific detail. Additionally, statistical details are often published in accordance with previously classified formats (industrial employment, etc.). Demographic data cannot be shown by linear devices, and similarly area symbols are inappropriate unless the available statistics are related to specific regions (e.g. maps illustrating population density). In the latter case, enumeration districts can be depicted by using a variety of tones of black, or a related sequence of colours, with each corresponding to a defined numerical class.

3.5.3 Linear representation: flowmaps

Geographical phenomena relating to a route, direction or movement are based on the employment of graphic techniques resulting in the generation of line symbols (rivers, roads, railway tracks, etc.). These form 'networks' which are graphic constructions comprised of straight line sections, of varying lengths, which join one with another. The purpose of a network is to demonstrate the interrelationships existing between various types of data distributed over the surface of the map. When considering the different networks, it is possible to discern diversities in terms of apparent importance (e.g. the road hierarchy), which are graphically demonstrated by noticeable variations in visual dominance, and in the amounts which are shown by changes in width (modifications of the thickness of line symbols) (see Fig. 3.58).

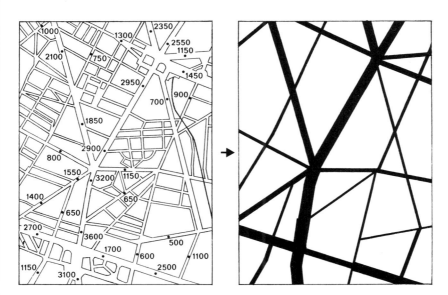

Fig. 3.58 Data relating to the number of cars passing per hour were collected between selected street intersections. Assuming a line width of 1 mm = 1 000 cars per hour, a network can be constructed with the displayed thickness of each road being made proportional to the number of vehicles recorded

The deliberate radial positioning of linear elements (as if to imitate the spokes of a wheel) pointing towards a centre, or the use of a series of concentric symbols based on a particular location, can provide a strong impression of 'attraction' or 'focusing' (migration, journeys to work, etc.) (see Fig. 3.38).

3.5.4 Areas

It has already been stated that in all cases relating to area representation:

- quantitative data should be grouped into classes; and
- an area symbol should cover the whole of a designated region in a uniform way.

Isopleths, isometric layers. When a feature on the land surface exhibits constantly changing values in all spatial directions (e.g. relief, temperature or precipitation), and has of necessity to be cartographically represented, it is normal that the only reliable quantitative data available have been collected at a few well-defined locations. These measurements are usually made with reference to spot heights or meteorological stations, etc. From these accurate observations, isopleths (lines joining points of equal value) are constructed to connect all appropriate points occurring within the mapped area, whether measured or interpolated, with an equivalent value. The isopleths define the boundaries between the various quantity classes. Each of the areas bounded by consecutive lines contains a range of quantitative

Fig. 3.59 Initially, 'round figures' (or characteristic) values, eventually to be connected by isopleths, are selected. Subsequently, the spot values, between which one or more isopleths will pass, are connected with straight lines. Assuming that amounts vary consistently between the recorded spot values, the points at which the isopleths will cut these lines can be interpolated. Finally, all plotted points of the same value are joined to create a series of curved isopleths

values occurring between the stated isopleth designations.

Graphic representation: a tone of black, or a variation in colour, is selected to correspond to each of the constituent classes within a graded series (see Fig. 3.59).

Alternate band maps. This system of representation enables demonstration of the coexistence of different data, or several components of the same variable, within a given territory or group of areas (aspects of land use, ethnic groups, socioprofessional categories, age groups, etc.).

In this case the map demonstrates two levels of information:

- the structure of data relating to each included area, i.e. the percentages of various components; and
- data distributed over the whole of a territory mapped, thus facilitating comparisons between its constituent areas (see Fig. 3.60).

The area being studied is overlaid with a series of parallel bands of constant width. Each of these represents 100%, and is further subdivided into strips with a thickness proportional to the occurrence of each component. The latter are illustrated by means of individually assigned colours or textures (see Fig. 3.61).

This technique only works well when there are distinct differences in percentages from area to area. When the variations are small, and consequently the graphic system does not show sufficient contrast, the alternate bands can be made proportional to 'frequency classes'.

Example: the colour or texture of a component will appear in one subdivision of the main band (divided in this case into strips with a constant width and value) only when it occurs in the corresponding percentage class; it will occupy two strips in a higher class; three in an even higher class, etc. The number of frequency classes should be equivalent to the number of strips contained

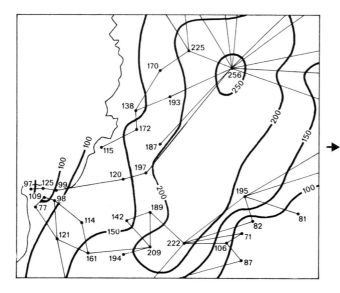

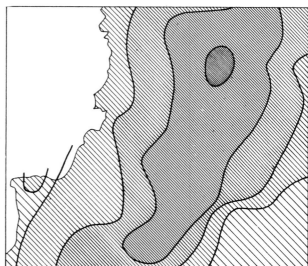

in one main band (see Fig. 3.62).

Alternate band maps have many other potential applications; for example, the illustration of standard deviations relating to different components of a variable. Their use enables the cartographer to overcome complex data display problems, and to represent detail on the map in a clear and efficient way.

Grid net maps. The technique involving the use of a grid was first developed several years ago in an attempt to overcome the problem of representing data relating to areas which vary significantly in size. Typically, these occur in large urban agglomerations and are very difficult to compare.

A uniform grid comprised of squares, rectangles or triangles is superimposed upon the whole of the territory being considered. This is thus subdivided into small units, of equal area, to which relevant statistical detail can be attributed. It is possible to depict one or more items of data referring to each of the grid units. Relative or absolute values can be demonstrated by the use of graded textures, colours, proportional point symbols, or a combination of these. Examples of the employment of this technique can be seen in the *Atlas of London*.

The limits of the system are reached when, as a result of the attempted combination of too many items of data, information becomes too complex and not immediately perceivable.

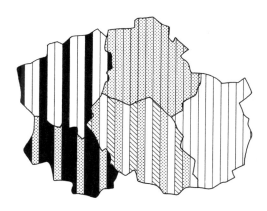

Fig. 3.62 Representation of 'dominant factors' based on frequency classes. In this case all areas contain three bands of constant width (each representing 100%) which are further divided into strips whose combined thickness is proportional to the value of each of the three components illustrated. If one of these has a value of between 40 and 50%, its assigned pattern or colour will fill one strip; between 50 and 60%, two strips; more than 60%, all three strips

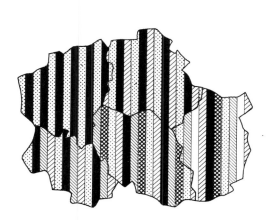

Fig. 3.60 The figure illustrates the coexistence of two, three . . . six data elements which are represented by parallel bands comprised of alternate strips of regularly arranged colours or patterns, all of a constant width. Within each area the composition of the band sequence includes as many differently coloured or patterned strips as are necessary to display all of the relevant data types. The band system is superimposed over the whole of the mapped territory

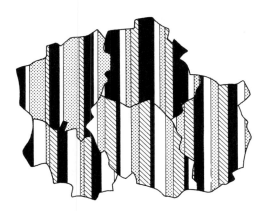

Fig. 3.61 The display of four components according to their percentage occurrence in each area. The total width of a unit (100%) is constant and is always divided into four bands. The thickness of each of the latter is proportional to the percentage of the component represented

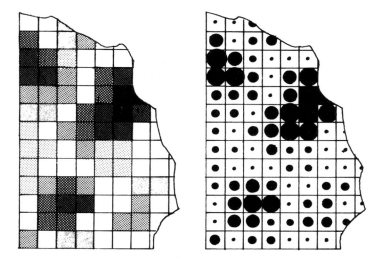

Fig. 3.63 Grid net maps consisting of small units of equal area. (a) Variations in value are shown by graded textures. (b) The same data are represented by proportional point symbols

Chapter 4

MAP DRAWING AND LETTERING TECHNIQUES

K. Kanazawa

CONTENTS

95

Standard symbols used throughout this chapter:

☺ Good ☺ Doubtful ☹ Poor or Inadequate

4.1 Introduction

Map drawing and lettering are fundamental techniques employed in the preparation and creation of fair drawings. In this chapter the application of traditional, basic tools in the generation of linework and lettering is described, and an explanation of operations and important procedures is provided. The principal aspects of processes are detailed and illustrated. Reference is also made to the main instruments, materials and methods which have most recently been introduced.

The chapter consists of eight sections, the first of which consists of general remarks relating to processes (i.e. their types and purposes), flow diagrams and the registration of materials. Section 4.2 provides information on basic equipment, instruments, materials and their usage. This is followed by pages explaining the principles and techniques of drawing with ink, i.e. methods for and rules relating to the construction of basic shapes and common symbols; together with details concerning the concepts involved in producing hachures and hill shading. Sections 4.4 and 4.5 introduce scribing and associated processes such as masking and stick-up. Lettering provides the topic for examination in Section 4.6 which considers basic theories and methods. The freehand creation of Roman alphabets is exemplified, and the equipment available for the mechanical generation of characters is described. Name placement on maps is illustrated in Chapter 2, section 2.27 of *Basic Cartography*, Volume 2. Section 4.7 focuses on lettering used to communicate in local languages and applied in mapping issued for domestic use, with the essential concepts of Sino-Japanese characters being employed as an example. Principles relating to bilingual and multilingual presentations using various lettering styles are explored; and photo-typesetting and computer-assisted type production are also discussed.

The rewritten version of this chapter incorporates major revision of many of the pages which appeared in the first edition, and the included illustrations have been described in detail as an aid to their improved understanding.

4.1.1 The purpose of map drawing — general remarks

Fair drawings are produced from map compilations which are generated as a result of research and the processing of available source materials. At this final stage of map preparation each of the included symbols must be produced according to the definitive instructions detailed on a specification list. In conventional reproduction, fair drawings will normally serve as original documents in the manufacture of plates for use in the photo-lithographic process, which allows the eventual printing of multiple copies. However, sometimes, particularly if only a small quantity of maps is required, finished drawings may be produced specifically for use with diazo or electrostatic copiers. In addition to these now 'traditional' techniques, computer-assisted methods, such as those constituting part of desk-top mapping systems, are now being developed.

The efficient communication of spatial information by means of a map is completely dependent on the quality of the graphic, i.e. it must be clear, legible, accurate, carefully produced, and not open to misinterpretation or misunderstanding whilst also creating the desired impression. These factors can be enhanced or destroyed by the quality of the employed draftsmanship. Thus linework must be precise in terms of its width, length and crispness; exhibit regular shape, orientation, intersections and angularity at corners; be of an appropriate density, sharpness and regularity; and allow the reader to distinguish between variations in width, most particularly where lines of different thickness connect one with another.

4.1.2 Methods of map drawing

4.1.2.1 Pencil or ink drawing on an opaque or translucent base material
Pencil linework is normally employed during the preparation of original survey plots; in the origination of very large-scale documents such as engineering drawings; for producing compilations or sketch maps; and in the generation of shading. Drawing pens and ink are used when particularly

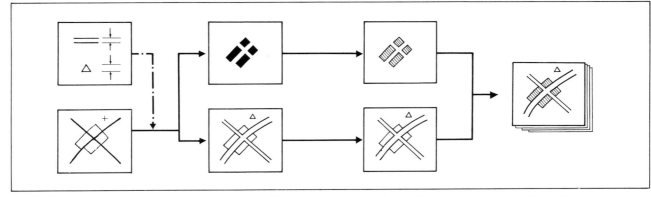

Fig. 4.1 Map component preparation on base materials using various drawing methods and according with definite instructions detailed on a symbol specification list. The results are usually produced as combined, multi-colour copies

Fig. 4.2 Precision of linework width, length and edges

Fig. 4.3 Regularity of shape, orientation and angularity

Fig. 4.4 Density of linework

Fig. 4.5 Sharpness of linework

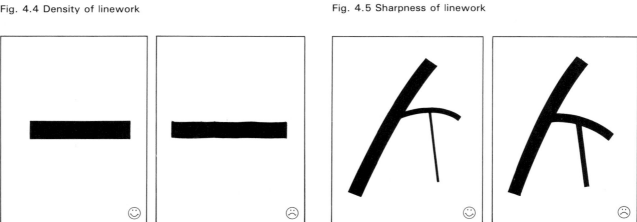

Fig. 4.6 Regularity of linework

Fig. 4.7 Accentuation of different linework widths and crispness of their inter-connections

Figs 4.2—4.7 Essential drawing qualities

98

clear and well-defined linework is required. The production of this is simpler and results in graphic detail which is fit for immediate use, but it requires considerable expertise. The application of tubular/reservoir pens with interchangeable points of varying thickness, or of technical pens, does not require the same amount of skill. Fair drawing of a significant amount of complicated and fine linework, such as that appearing on a topographic map, is normally undertaken at a scale larger than that of the intended end-product, i.e. 125%, 150% or 200%. A process camera may be used to enlarge source detail to the working scale, and subsequently to reduce the drawn imagery to the final size and convert it to a negative original for use in the manufacture of a printing plate. The fair drawing of simpler linework, for example the content of an engineering plan or the outline of an area symbol on a choropleth map, can be produced at same-size by tracing detail onto a sheet of tracing paper or plastic film superimposed upon a compilation.

4.1.2.2 Scribing and mask making
Scribing involves the engraving of linework on the actinically opaque coating of a transparent, stable-based plastic film in order to produce a negative image. Its main advantage is that detail of a consistently high quality can be generated quickly and easily, at the required scale, without the need to use a camera. The choice as to whether to scribe or draw linework in ink depends on the availability of materials and equipment, and the expertise of staff. There are two methods used for the creation of a scribed image — 'wrong reading' (mirror image) and 'right reading'. The latter is the simpler to check against a compilation, but its use necessitates the employment of a special process during the manufacture of offset printing plates.

Very fine area symbols or patterns are produced by making an 'open window' in a sheet of peelcoat. This is done by cutting and removing an opaque membrane from the surface of a base material. The resultant negative mask is then photographically exposed in association with an appropriate negative screen tint, during contact copying, to produce a toned, final positive original.

4.1.2.3 Painting, stick-up and dry-transfer
Painting may be carried out by using either a brush or an airbrush. It is employed, rather than peelcoat, to produce masks for large areas destined to appear either as even tints, tints decreasing in density (vignettes), or as an overall grey shading providing coverage of the complete area mapped. The size of brush used depends on both the extent of an area and the complexity of its outline.

Stick-up (applied with wax or a solvent) and dry-transfer (which involves peeling from a backing sheet) are both terms applied to the removal of an image from a ready-made sheet of identical symbols (i.e. lettering, settlement symbols, etc.), and positioning it on an original drawing. The sheets may be commercially produced or made in the drawing office using a phototypesetter or a contact copier.

4.1.3 Map drawing flow diagrams

Preparation of the many separate drawings required for each of the colours that it is intended should eventually be printed, to produce a multi-colour map, involves complex procedures demanding the making of decisions at every stage. This can best be illustrated by means of a flow diagram incorporating symbols explaining the nature and use of employed materials, together with the processes undertaken. Careful consideration, and the making of judgements relating to the organisation of map preparation, is normally necessary before starting work. Compilation of a flow diagram is strongly advised even in comparatively straightforward cases. Besides detailing the stages of photomechanical working, it can be further expanded to include a work schedule, estimates of required manpower and levels of expertise, and the costs of employed materials.

Figures 4.11 and 4.12 are examples of flow diagrams including graphic symbols. The first relates to a right-reading, pen and ink drawing method, whilst the latter demonstrates the stages necessary to produce the same three-colour map using a wrong-reading scribing technique. Thus a visual comparison of the two approaches can be made. The steps relevant to drawing procedures are described in depth, but only the most essential are included concerning compilation, platemaking and printing. Photographic enlargement and reduction work, which is normally associated with pen and ink drawing, has been deliberately omitted in order not to make the diagram unnecessarily complicated.

A key to the graphic symbols is provided, and each of the squares constituting the flow diagram represents a sheet of the relevant material used at each stage. The characteristics of resultant images are described in the four corners of each square, i.e. ① image position at the top left; ② image form at the bottom right and diagonally opposite ①; ③ image type at the top right; and ④ method of image formation at the bottom left. If necessary, the type of material used can also be symbolised at ④. The various stages are demonstrated by a number in the centre of each square, and four different sorts of connecting lines are included in order to demonstrate the sequential relationship of the stages. Types of drawing procedures employed are illustrated by arrow-heads of different characters, which are positioned at the end of the lines joining the squares. Photo-

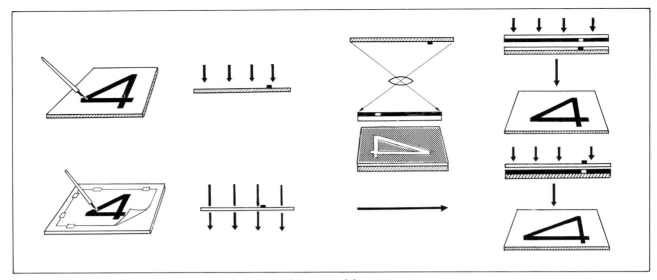

Fig. 4.8 Pencil or ink use on an opaque or translucent base material

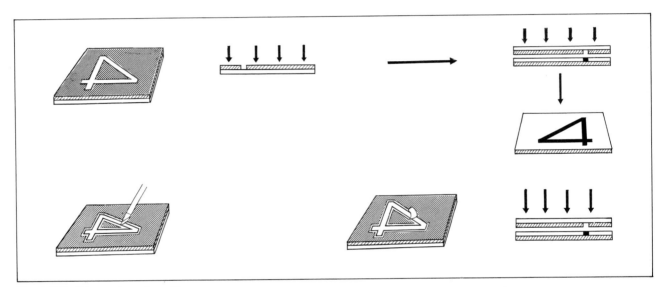

Fig. 4.9 Scribing and mask making

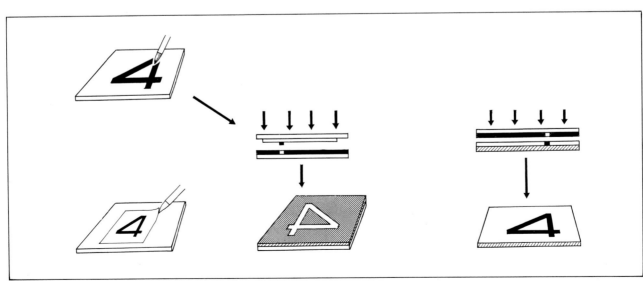

Fig. 4.10 Painting, stick-up and dry-transfer

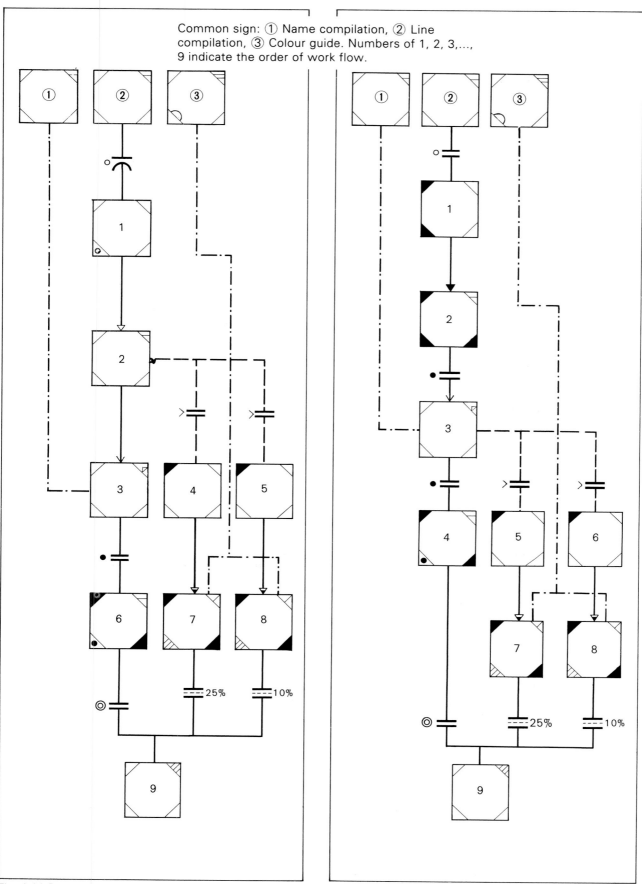

Common sign: ① Name compilation, ② Line compilation, ③ Colour guide. Numbers of 1, 2, 3,..., 9 indicate the order of work flow.

Fig. 4.11 Pen and ink drawing
Standard symbols: 1 Name compilation; 2 linework compilation; 3 colour guide. The numbers 1, 2, 3 . . . 9 indicate the sequence of working operations

Fig. 4.12 Scribing
Standard symbols: 1 Name compilation; 2 linework compilation; 3 colour guide. The numbers 1, 2, 3 . . . 9 indicate the sequence of working operations

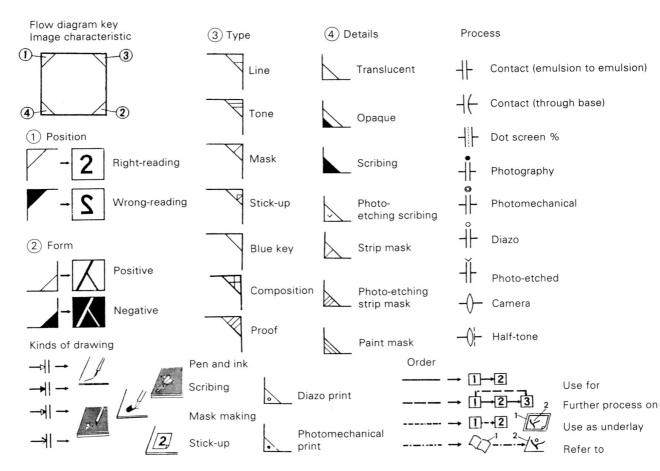

Key to Figs 4.11 and 4.12

graphic or non-photographic processes are indicated by the inclusion of specific symbols half-way along these lines.

In Figs 4.11 and 4.12, ①, ② and ③ included in the top row of squares relate to elements of the compilation process such as: name forms, type styles and individual point symbols; line drawing of shading, etc., usually in pencil; and a colour guide to area symbols which may often be provided by painting a corner of the square in the specified hue. The final square, indicated by the number 9, represents a positive image formed by the non-photographic overprinting of detail using the three colour separation drawings. This normally serves as a proof sheet, but may sometimes be the end product. After the making of any corrections required, the negative originals, demonstrated by the figures 6, 7 and 8 in both flow diagrams, are sent to the print room and used in the manufacture of lithographic plates.

4.1.4 Register marks and punch-registration systems

Modern map drafting normally involves the preparation of separate overlays for individual components such as linework, area masks, lettering and colour elements. To ensure that all of these fit together correctly at the printing stage, it is essential to employ register marks on each of the individual sheets. These enable the accurate fitting together of the various composing documents. They are used throughout the production phase from initial drawing, through photomechanical processing and proofing, to the ultimate printing of the map. Normally the marks consist of crosses, or refinements of these, which are either drawn in ink or stuck down on positive materials, or scribed on negative scribecoat. It is usual to position them in each of the corners of the sheet, outside the working area.

As individual map elements often occur in a negative form, and as photomechanical operations normally take place in a darkroom, it is extremely difficult to see and register these marks accurately one on top of another. To overcome this, several systems of punch-registration have been developed and consist of instruments designed to cut circular holes or slots in film or plastic drawing foils. These make it possible to pre-punch sets of repromat materials and facilitate their positioning over individual studs, or a pin bar, in any combination and in perfect register.

Systems range from simple two- or three-hole punching devices, used for small format work, to elaborate five- or seven-hole generation units that can accommodate large size base materials. In both cases it is necessary that the holes are

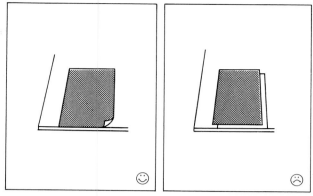

Fig. 4.13 Exact register between all sheets combined to produce a multi-colour map is essential

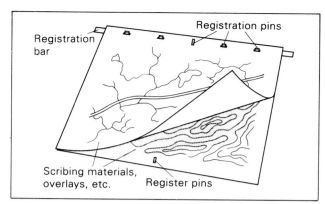

Fig. 4.14 The principle of punch-registration. Holes are punched in all sheets which are then mounted on a pin bar or studs

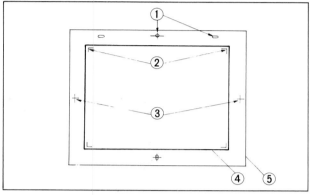

Fig. 4.15 Positioning of punch holes and register marks. 1 Holes; 2 corner marks; 3 side marks; 4 neat line; 5 sheet

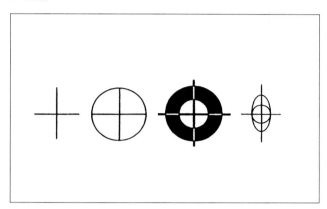

Fig. 4.16 Examples of register marks.

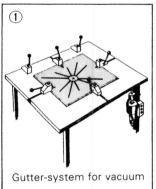

Fig. 4.17 Capabilities of different punching systems. 1 Four sides; 2 one side; 3 two sides

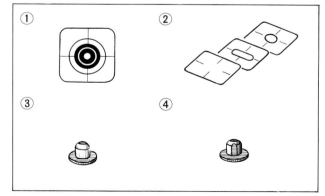

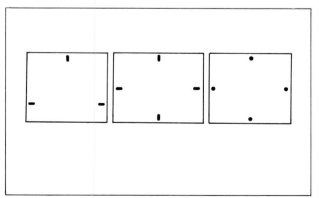

Fig. 4.18 Examples of punch hole positions on a sheet

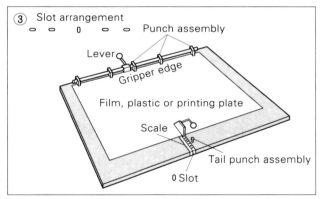

Fig. 4.19 Registration aids. 1 and 2 Film originals for contact printing, register marks on sheets; 3 pin bar stud; 4 stud

103

punched, at fixed distances, sufficiently far apart to prevent the material moving about. When working with large sheets it is desirable to employ a register system capable of producing an additional slot in the centre of the side of the material which is opposite the gripper edge. This serves to provide extra control.

4.2 Instruments, tools and materials

4.2.1 The layout and lighting of a drawing office

4.2.1.1 Workstations
Recommended dimensions, in millimetres, for a person with an average height of 1 651 mm:
①　1 285; ②　800 or more; ③　700; ④　1 183; ⑤　600; ⑥　200; ⑦　400; ⑧　800; ⑨　500; ⑩ 700; ⑪　1 200; ⑫　700; ⑬　1 651; ⑭　25; 15　1 556; 16　1 600 to the top of the drawing board.

4.2.1.2 Lighting
① Overall office average of 500 lumens emitted by fluorescent tubes; ② drawing board illuminated by white fluorescent tubes or an adjustable incandescent lamp which can be moved to eliminate shadows; ③ 700−1 500 lumens shining onto the board; ④ height at which overall illumination is measured; ⑤ 850 mm.

4.2.1.3 Spacing between workstations
①−④ Without a gangway; ⑤−⑧ with a gangway.

4.2.1.4 Drawing desk
① It is desirable to work at an inclined drawing board; ② an adjustable angle drawing board.

4.2.1.5 Desktop graphics workstation
Increasingly computerised workstations are becoming a feature of cartographic laboratories. Operators employ these for the interactive preparation of compilations and manuscript materials by using a keyboard and a 'mouse' to manipulate displayed data.

4.2.1.6 Light table
It is advisable that illumination of the table should be provided by white light of between 200 and 1 000 lumens intensity as compared to ambient lighting in a ratio of 10 to 6 or 9. Uniform illumination, together with some protection from the heat emitted by the lamps, can be provided by incorporating: ① sheets of 5 mm thick glass which is matt on one side; ② a sheet of milky-white plastic; ③ opal glass. The light source should consist of either 20 W fluorescent or daylight tubes spaced at 20-cm intervals.

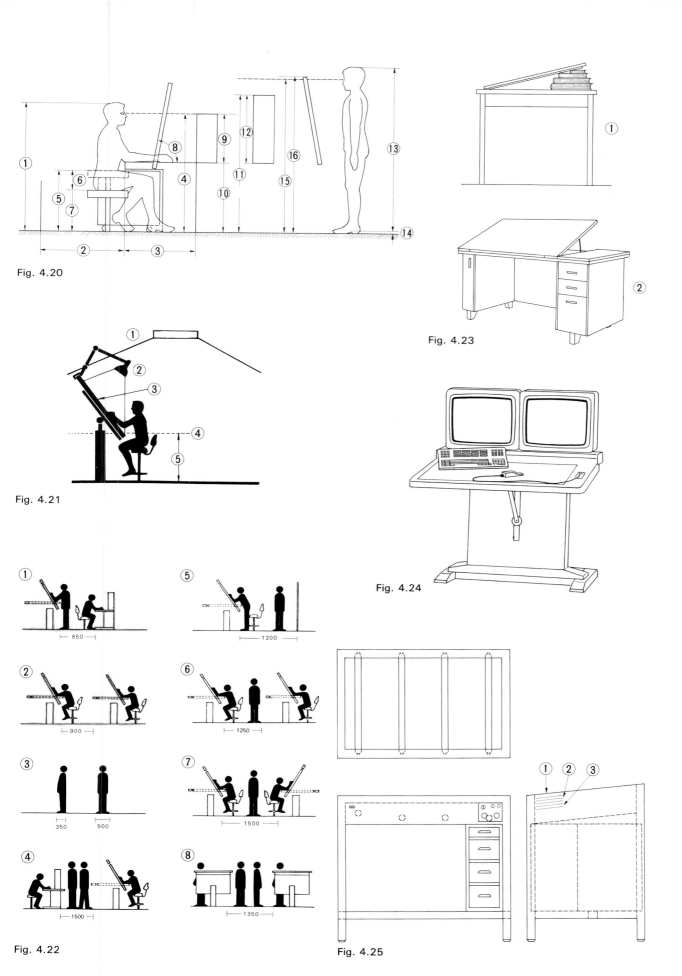

Fig. 4.20

Fig. 4.21

Fig. 4.22

Fig. 4.23

Fig. 4.24

Fig. 4.25

105

4.2.2 Instruments and tools for use in compilation

4.2.2.1 Proportional dividers

These can be useful in the transferring of detail at a different scale to that of an original. The adjustable fulcrum ① is set to the appropriate ratio scale which is marked on the face of the dividers. A distance is then measured off from the map. The points at the other end of the instrument demonstrate the requisite amount of reduction or enlargement.

4.2.2.2 Scale or projection transformation using grid squares

Details from the original map document ① are copied, square by square, onto the second grid which consists of the same number of squares ②.

The pantograph: although now somewhat dated, this instrument is still useful for the reduction or enlargement of detail by up to 6 or 7 times. Additionally, it can be employed in the copying of drawings consisting essentially of a complex of curved lines.

4.2.2.3 Simple pantograph

A required ratio or proportion must be pre-specified. The stylus ① and pencil ② are interchangeable for use during reduction or enlargement. ③ This is the pivotal point.

4.2.2.4 A pantograph assembled to allow reduction

If K is the reduction ratio, X the length of an arm, and Y the adjusted length for moving the arm and pencil position, then $Y = KX$. For example: if $X = 60$ cm, $K = 1:25\ 000/1:10\ 000$, then $Y = 60/2.5 = 24$ cm.

Adjustments should be made to ensure that the stylus ①, pencil ②, and pivotal point ③ are kept in a straight line.

4.2.2.5 A pantograph assembled for enlargement

In this case $Y = KX (1 + K)$. For example: if $X = 60$ cm, $K = 1:10\ 000/1:25\ 000$, then $Y = 2.5 \times 60 (1 + 2.5) = 42.9$ cm.

4.2.2.6 Principles relating to different types of optical pantographs

① Original map; ② image; ③ lens; ④ mirror.

4.2.2.7 Dividers

A 'pair' of dividers is an instrument enabling the transfer of a specific distance measurement from one place to another, or the multiple extension of a given length. In the latter case, the sequence of operations is shown in Fig. 4.32 by the numbers ①, ② and ③.

4.2.2.8 Design/spacing dividers

Shown in Fig. 4.33.

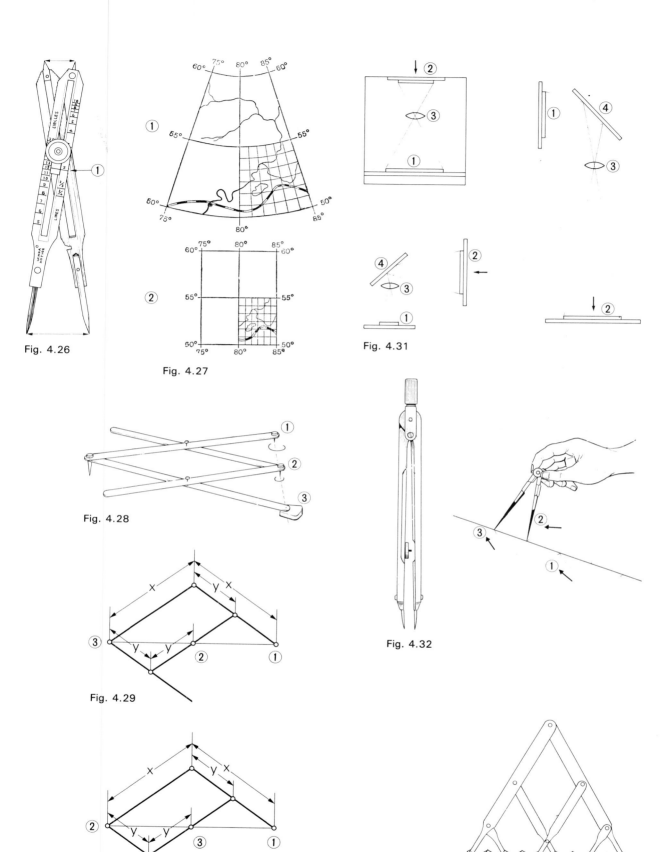

Fig. 4.26

Fig. 4.27

Fig. 4.28

Fig. 4.29

Fig. 4.30

Fig. 4.31

Fig. 4.32

Fig. 4.33

107

4.2.3 Rulers, straight-edges, set-squares and protractors

4.2.3.1 Rulers and straight-edges

The former are accurately calibrated and essentially used for the measurement of linear distances; in addition, they can be employed in the construction of lines of predetermined lengths. A straight-edge is specifically intended to aid in the drafting of straight lines. The trueness of its edge can be tested by drawing a fine line along the entire working length. After completing this, the instrument is inverted and the procedure repeated by redrawing the original line using the same edge. Imperfections will be immediately noticed, and can be corrected by carefully scraping or sanding if the straight-edge is manufactured from wood or plastic.

4.2.3.2 Set-squares

Figure 4.35(a). Checking the accuracy of a 90° (right angle) corner.
Figure 4.35(b). Checking the accuracy of 60° and 30° corners ①. Checking the accuracy of 45° corners ②.
Figure 4.35(c). ① Using a pencil with a specially processed straight-edge; ② using a processed edge with a technical pen, ruling pen, etc.

4.2.3.3 Using a T-square

The construction of vertical or inclined parallel lines using a T-square in combination with a set-square.

4.2.3.4 Protractors

① Semi-circular (180°) protractor; ② circular (360°) protractor incorporating a precise angle-setting device.

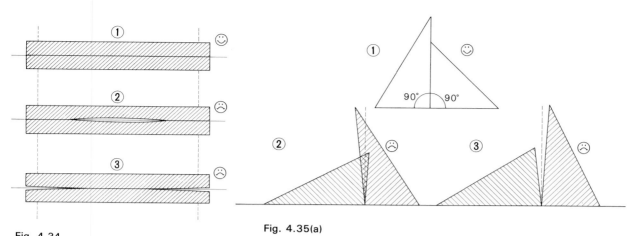

Fig. 4.34

Fig. 4.35(a)

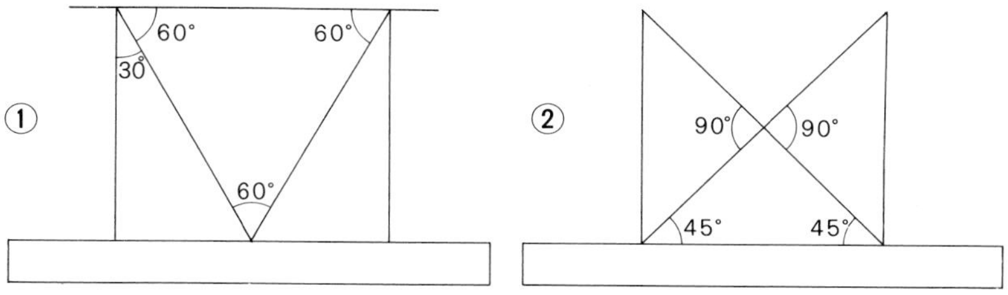

Fig. 4.35(b)

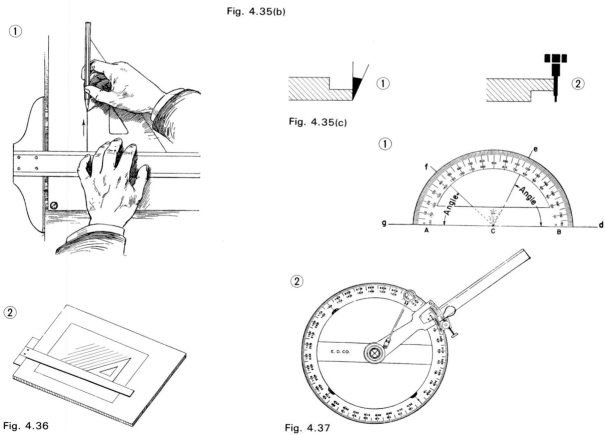

Fig. 4.35(c)

Fig. 4.36

Fig. 4.37

4.2.4 Constructing hatching and curved linework

4.2.4.1 Producing hatching by using straight-edges and a set-square
① Set-square with two sides of equal length (an isosceles triangle); ② a specially manufactured straight-edge; ③ a second straight-edge positioned at an angle of 45° to the horizontal; ④ paperweights; ⑤ the interval relating to the required space between the lines drawn; ⑥ resultant lines comprising hatching.

4.2.4.2 Manually operated cross-hatching machine
1 Operating button; 2 straight edge which slides down by a predetermined amount when the button is pressed; 3 Vernier magnifying scale for checking interval spacing; 4 a dial for the adjustment of interval spacing.

4.2.4.3 Fixed position drafting machine used for hatching
① Supporting bar; ② position fixing points for the supporting bar; ③ a movable straight-edge allowing the spacing interval to be adjusted by means of a dial.

4.2.4.4 Glass straight-edge
Constructed from a cylindrical glass rod, this instrument has a diameter of 9−10 mm ①, a length of 25−30 cm ②, and incorporates two collars made of vinyl, rubber or paper 7−8 mm wide and 0.5 mm thick ③. It is useful for drawing short straight or slightly curved lines measuring 5−6 cm or less.

4.2.4.5 Using a set-square as a straight-edge
Note the method of holding and the thickness of the set-square.

4.2.4.6 Templates for drawing circles
① Elliptical templates; ② bow templates.

4.2.4.7 Flexible curve and tubular pointed pen
Shown in Fig. 4.44.

4.2.4.8 Spline curve and weights
Shown in Fig. 4.45.

4.2.4.9 French curves
Shown in Fig. 4.46.

Note: A smooth curve can be accurately constructed by using the equipment described in 4.2.4.8 or 4.2.4.9. It is necessary to select at least three points along the desired curve and then to join these using an appropriate edge of the spline. A line is then drawn connecting the first two points, and subsequently the construction is extended to incorporate the others.

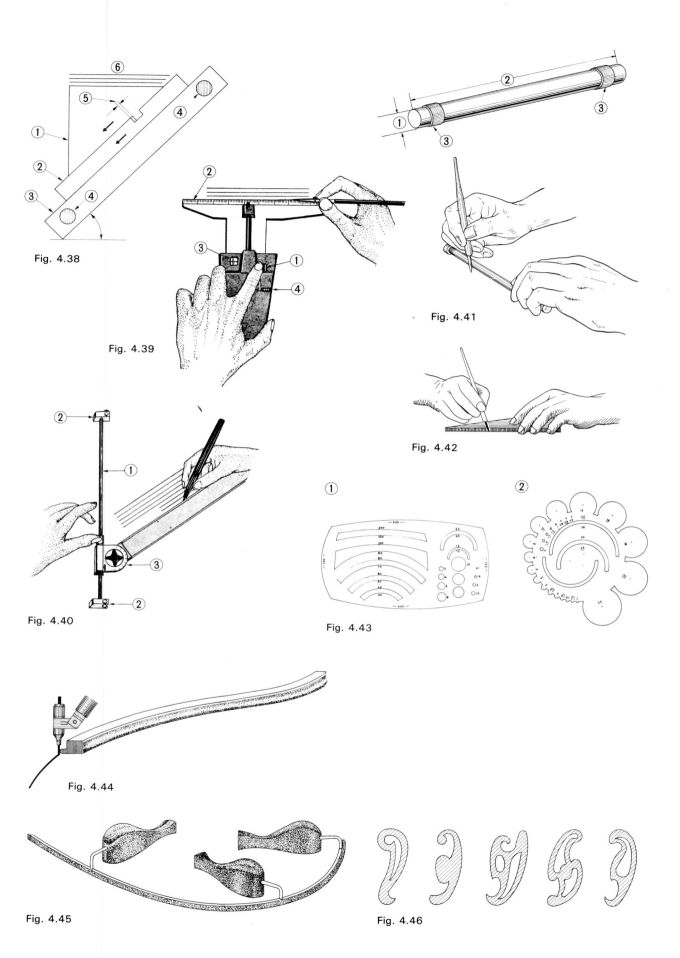

Fig. 4.38

Fig. 4.39

Fig. 4.40

Fig. 4.41

Fig. 4.42

Fig. 4.43

Fig. 4.44

Fig. 4.45

Fig. 4.46

111

4.2.5 Pencils and erasers

Note: It is unadvisable to use a cylindrically shaped pencil because these are difficult to grip firmly and also tend to roll off a drawing board. Lines produced in pencil should be drawn firmly, be sharp, and even in thickness. The latter quality can be achieved by twisting the pencil in the fingers as a line is being drawn. The angle between the pencil and drawing surface should be maintained at about 80°. Practice is necessary to prevent denting the material, especially if it is paper. This is an essential skill in map drafting.

Black lead pencils: care should be taken to select an appropriate grade of pencil which exhibits the desired degree of hardness and blackness: 4H−3H: these are used to generate intricate detail on hard surfaced papers or plastic foils in dry weather conditions; 2H−H: employed on soft surfaced papers or other media in wet weather; HB−B: suitable for drawing on smooth, calendered tracing papers.

Crayons or coloured pencils: again appropriateness for particular use is determined with respect to quality: hard lead: these are equivalent to 2H black lead pencils and are used to colour intricate detail; soft lead: appropriate for use in shading areas; oily lead/'Chinagraph' pencils: suitable for producing drawings on smooth polyester foils, glass plates, film, calendered printing papers, etc.; watercolour: soluble leads which can be employed to produce a colour wash effect when used on drawing paper; proof reader's pencil: this incorporates a blue lead suitable for the noting of errors on detailed drawings. If subsequently photographed, light blue lines do not reproduce as part of the image if normal developers are used.

4.2.5.1 A pencil lead should be sharpened to produce a conical point for use in normal drafting
① 10 mm in length for H grade pencils, 15 mm for 4H, and 15 mm into the wooden lead holder.

4.2.5.2 The lead should be sharpened to produce a chisel point when intending to generate straight lines
① As for Fig. 4.47.

4.2.5.3 Lead intended for use in a compass should be sharpened as suggested in Fig. 4.48
Shown in Fig. 4.49.

4.2.5.4 Pointing a lead with sandpaper attached to a wooden block
① Sharpen in a horizontal direction and then ② vertically. Afterwards, rub the point on rough sandpaper to ensure sharpness.

4.2.5.5 Pencil pointers
① Coarse and fine graded surfaces together with a plastic sponge for cleaning dust from the lead. Both conical and chisel points can be produced; ② a pointer designed to produce conical points on clutch pencils.

4.2.5.6 Clutch pencils
① Sliding sleeve type; ② push button variety.

4.2.5.7 Glass eraser
① Push button type; ② spare glass fibre; ③ sliding sleeve variety.

4.2.5.8 Indiarubber eraser
① Rubber; ② holder; ③ cleaning brush; ④ push button.

The example at the top of Fig. 4.54 is a pencil type with the rubber encased in wood; both of the lower examples are available with replacement rubbers.

4.2.5.9 Erasing shield
Normally made of steel, these templates facilitate the removal of unwanted detail from small or restricted areas.

4.2.5.10 Draftsman's brush
This is used to clean a drawing board or table.

4.2.5.11 Various types of electric erasers
These may be either battery or electrically powered. ① Rubber; ② on/off switch.

4.2.5.12 Erasing knives
The orientation of the blade should be modified to meet the drawing. Erasure of unwanted detail is carried out by softly scraping the surface with the knife held at an angle (α).

4.2.5.13 The point of the knife
① Point; ② plan view; ③ side.

4.2.5.14 Erasers used in the removal of small areas of detail from tracing paper or plastic foil
① A commercially manufactured cutting knife or 'lance eraser'; ② a knife produced by sharpening the nib of a drawing pen to produce an angle. These are used to scrape the surface vertically.

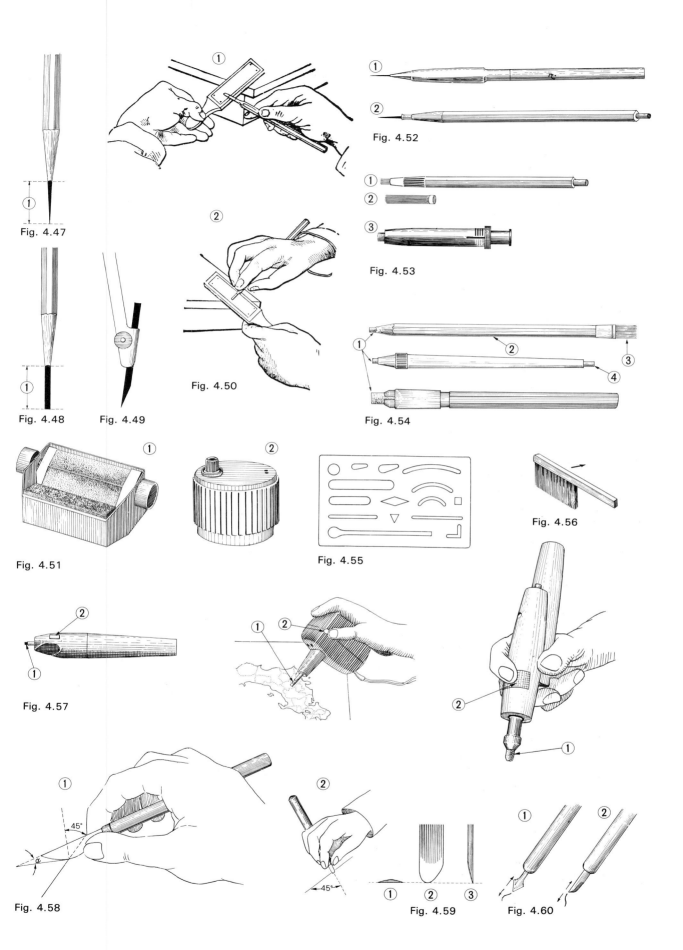

Fig. 4.47

Fig. 4.48   Fig. 4.49

Fig. 4.50

Fig. 4.51

Fig. 4.52

Fig. 4.53

Fig. 4.54

Fig. 4.55

Fig. 4.56

Fig. 4.57

Fig. 4.58

Fig. 4.59

Fig. 4.60

113

4.2.6 Drawing pens

4.2.6.1 The drawing/mapping pen, sometimes known as a 'crowquill'

This is the primary tool used in pen and ink drafting. Figure 4.61 illustrates a nib with a cylindrical/sleeve base which fits directly into a penholder. ① Rear view; ② front view; ③ side view.

4.2.6.2 Drawing nib with a curved base

Nibs of this type are normally longer than those described in 4.2.6.1, and are particularly used for lettering. ① Front view; ② front view of a sharpened nib; 3 side view.

4.2.6.3 Nib shapes

① The central slit is straight. Metal elements on either side of it are tapered and joined below the base of the slit; ② both sides of the nib are rounded and touch the drawing surface smoothly; 3 the base of the nib's tip is also rounded to ensure smooth contact with the surface. ④ – ⑦ These are bad examples: ④ after being in contact with the drawing surface the two sides do not close together properly; ⑤ the two sides of the nib are not level at the top; ⑥ neither of the sides is symmetrical with the central slit; ⑦ the slit is at an angle to the axis of the pen.

4.2.6.4 Shaping the nib

Both sides of the nib are shaped by using a grind-stone to increase the degree of tapering. If this process is not performed carefully, or if too much pressure is exerted, the top of the slit will be worn away and the nib will be of no further use.

4.2.6.5 Grinding the nib

① The outer edges of the nib are ground and sharpened several times; ② subsequently, the top corners of the nib are processed by rotating the pen, held vertically, 2 or 3 times on the grind-stone. The latter can be produced by mounting silicon-backed, waterproof emery-paper on a block of wood. Papers Nos 400–600 are available for coarse grinding, Nos 800–1 000 for fine grinding and finishing.

4.2.6.6 Shaping the outer edges of a nib

① The nib should be frequently inspected with the aid of a folding magnifier/linen tester, which gives 3–5 times enlargement; ② a cylindrical glass ruler, wrapped with emery-paper, may be used to assist when shaping the outer edges of a nib.

4.2.6.7 Penholders

The holding position adopted when using a drawing pen to create fine detail is lower on the shaft than that employed when working with a normal pen.

4.2.6.8 Holding a drawing pen

① When drawing fine detail, the pen should be slanted at an angle of about 70° to the board ② and ③. In the case of less complicated work the angle of the holder is inclined more from the vertical, and the point of grip is further away from the nib. ④ The axis of the penholder should always coincide with the drawing direction. ⑤ When working on fine detail, delicate strokes can be added by supporting the hand holding the pen with the forefinger of the other hand, whilst keeping the holder inclined at 60–70° to the surface. The drawing direction is at an angle of approximately 70° with the horizontal line of the desk. ⑥ If tracing on paper or linen, the material should be touched with the base of another pen-holder held in the opposite hand. ⑦ The quality of the drawing can be checked occasionally by inserting a piece of opaque white paper under the tracing medium from any side.

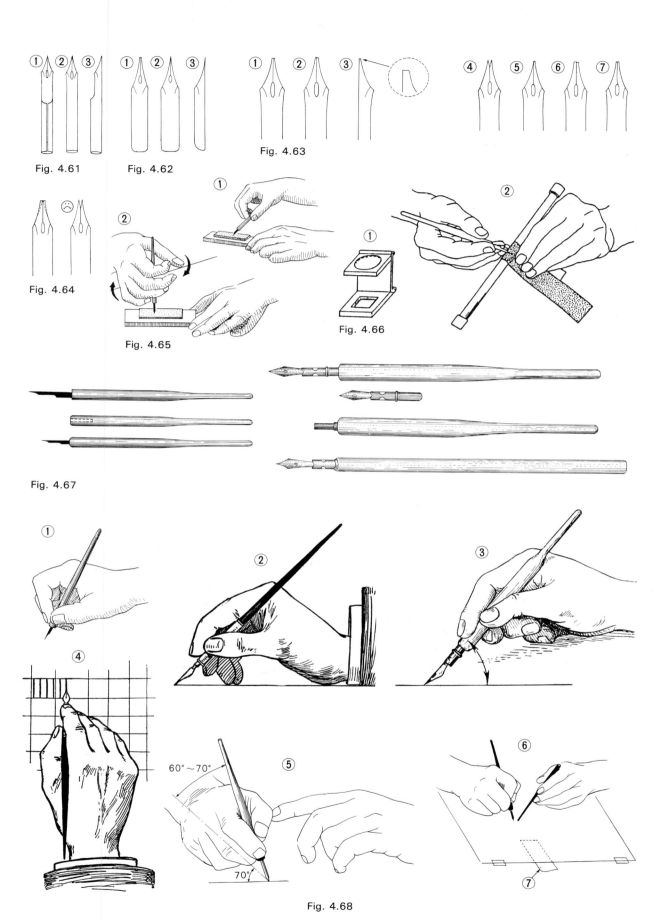

Fig. 4.61

Fig. 4.62

Fig. 4.63

Fig. 4.64

Fig. 4.65

Fig. 4.66

Fig. 4.67

60° ~ 70°

70°

Fig. 4.68

115

4.2.7 Ruling pens

4.2.7.1 Use of ruling pens
These instruments are specifically intended for use in the generation of straight or gradually curving lines of a predetermined and uniform width. Increasing the distance between the blades too much results in drawing difficulties. ① A fine-line ruling pen; ② standard model; ③ border pen; ④ three-bladed pen for thick lines; ⑤ double line ruling pen with an interior adjusting wheel; ⑥ double ruling pen with an external adjusting wheel. It should be noted that type ⑤ is more commonly available than ⑥.

4.2.7.2 Sharpening the tip of a ruling pen
① Plan view of the grinding process; ② side view. Sharpen gently whilst turning the blade slightly.

4.2.7.3 The shape of a point
① The point must exhibit a regular shape; ② – ④ it should be properly rounded and symmetrical with the centre line of the blade.

4.2.7.4 Care must be taken to avoid sharpening only part of the tip
Shown in Fig. 4.72.

4.2.7.5 The shapes of points as viewed from the side
① Border pen used to produce thick linework; ② fine-line ruling pen.

4.2.7.6 Filling a pen using a strip of paper
Shown in Fig. 4.74.

4.2.7.7 The pen should always be held at a right angle to the drawing surface
Shown in Fig. 4.75.

4.2.7.8 The blade of the pen should not rest against the side of a ruler
Shown in Fig. 4.76.

4.2.7.9 Typical errors occurring when generating lines with a ruling pen
① Too much pressure exerted against the edge of a ruler; ② blade sloping away from the edge of a ruler; ③ drawing too close to the edge of a ruler can result in ink running underneath it; ④ ink on the outside of the blade may run under the ruler; ⑤ pen blades not kept parallel to the ruler's edge; ⑥ ruler rubbed across the wet ink of a recently drawn line; ⑦ insufficient ink in the pen to enable completion of a line.

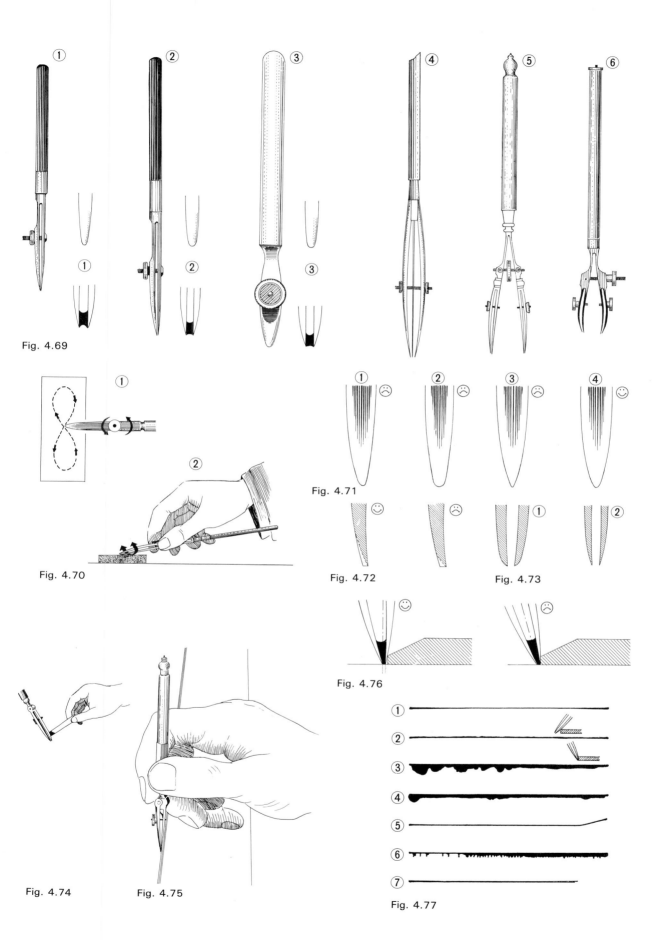

Fig. 4.69

Fig. 4.70

Fig. 4.71

Fig. 4.72

Fig. 4.73

Fig. 4.76

Fig. 4.74

Fig. 4.75

Fig. 4.77

4.2.8 Compasses

4.2.8.1 Beam compass
① Drawing element; ② beam; ③ pivotal point; ④ adjustment dial, ⑤ clamping screw, ⑥ pencil point; ⑦ ruling pen point.

4.2.8.2 Using a beam compass
Fit a needle point at the end of the bar to serve as a pivot. Determine the appropriate position of the drawing head on the beam by using the scale, and fix it by tightening the clamping screw. The exact location of the pencil point or tip of the ruling pen can be controlled by the adjustment dial. Carefully support the bar with both hands and turn it slowly and gently in order not to bend the beam.

4.2.8.3 Compass with interchangeable points and an extension arm
① Dividers; ② pencil point; ③ ruling pen point; ④ extension arm.

4.2.8.4 Using the instrument detailed in 4.2.8.3
Both the needle point and the drawing point must be adjusted to meet the surface at a vertical angle.

4.2.8.5 Spring bow compass
① Internal adjustment screw; ② draw circles by rotating the head of the compass with the thumb and forefinger.

4.2.8.6 Drop compass
This instrument is specifically designed for the construction of very small circles. The centre needle is positioned at an angle vertical to the drawing surface, and the top is held by the forefinger. A circle is then drawn by spinning the head using the thumb and middle finger.

4.2.8.7 Circle construction using a compass
①, ② and ③ illustrate the drawing technique.

4.2.8.8 Drawing concentric circles
Shown in Fig. 4.85.

4.2.8.9 Points of contact between circles should be precisely tangential
Shown in Fig. 4.86.

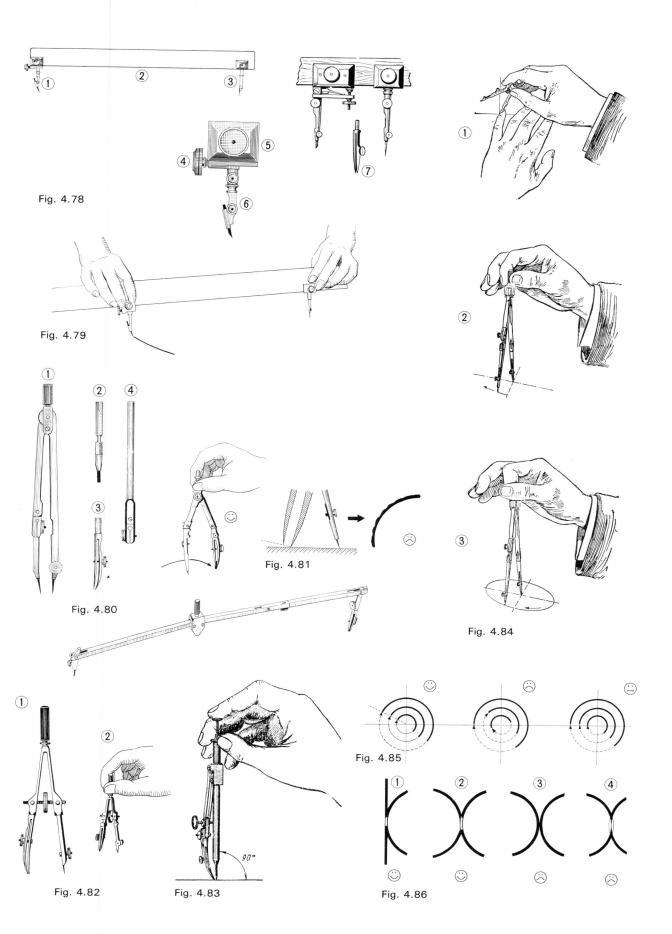

Fig. 4.78

Fig. 4.79

Fig. 4.80

Fig. 4.81

Fig. 4.84

Fig. 4.82

Fig. 4.83

Fig. 4.85

Fig. 4.86

90°

4.2.9 Contour and double ruling/road pens

4.2.9.1   Standard contour pen
The swivel axis is short and movement is easy.
① Eccentricity—the amount of offset of the drawing point from the central axis.

4.2.9.2   If necessary, the tip can be sharpened using the technique described in 4.2.7
Shown in Fig. 4.88.

4.2.9.3   Holding a contour pen
Shown in Fig. 4.89.

4.2.9.4   Contour pen with a long swivel axis
The axis can be clamped by using the adjustment screw at the top of the handle, and the instrument can then be used as a ruling pen. The degree of eccentricity is smaller than that of a normal type, and the pen can be used to draw slightly curved contour lines.

4.2.9.5   Holding a contour pen with a long swivel axis
Shown in Fig. 4.91.

4.2.9.6   Drawing is easily accomplished by working from bottom left to top right at all times
Shown in Fig. 4.92.

4.2.9.7   Use of the contour pen
① A join should always be made, if necessary, at the centre of the smallest curve; ② – ④ attempting to connect lines at other points provides unacceptable results.

4.2.9.8   Gaps occurring in a pecked line
These should be located on gentle curves rather than at the apex of sharp curves or corners. ①, ② and ④ illustrate good practice, but ③ and ⑤ are examples of bad working.

4.2.9.9   Double ruling pen or road pen
Shown in Fig. 4.95.

4.2.9.10   The method of holding a double ruling pen
This is the same as that adopted with a contour pen.

4.2.9.11   The tips of both pens
Both tips of each pen should rest on the base material as is shown in ①; ② demonstrates incorrect usage.

4.2.9.12   A double line drawn with a road pen but with sharp curves omitted
These are subsequently added, using a drawing pen and a ruler, and connect with the linework produced by the double ruling pen (see ① – ③).

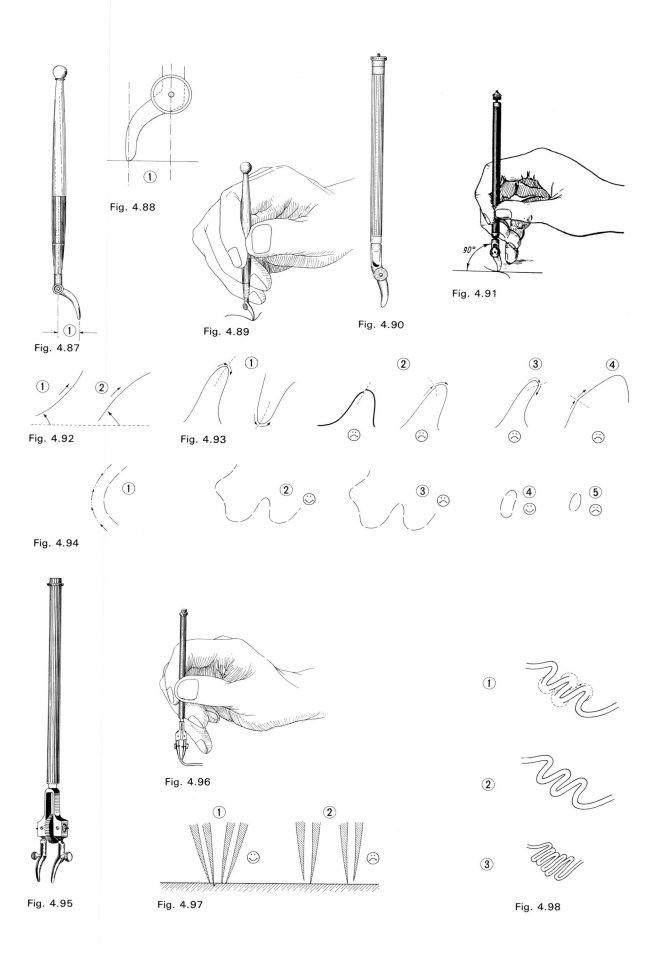

Fig. 4.88

Fig. 4.87

Fig. 4.89

Fig. 4.90

Fig. 4.91

Fig. 4.92

Fig. 4.93

Fig. 4.94

Fig. 4.95

Fig. 4.96

Fig. 4.97

Fig. 4.98

4.2.10 Special drawing pens and inks

4.2.10.1   Various sizes of flat-ended nibs
Shown in Fig. 4.99.

4.2.10.2   Double-ended nib
Shown in Fig. 4.100.

4.2.10.3   Round-ended nib
Shown in Fig. 4.101.

4.2.10.4   Tubular or reservoir pen
① Leroy pen; ② tubular point held vertical to the drawing surface and tangential to a straight-edge. The main advantage of this type of instrument is that its interchangeable points are especially designed to allow the free flow of ink and the generation of linework of constant thickness. Points range in width from 0.1 mm upwards, so providing considerable line−weight variety. Pens of this type can be used for drawing curves or for the production of lettering using templates. However, they do exhibit certain disadvantages. For example, the edges of lines constructed using these instruments are not as sharp as those produced using a ruling pen; there is a tendency for linework to demonstrate variations in width when points of 0.2 mm or less are used, unless it is drawn slowly. Pen manufacturers normally recommend an appropriate ink for use with their products.

4.2.10.5   Using a tubular or reservoir pen
① Good and bad examples of linework (from top to bottom): (a) a line drawn at a constant speed and demonstrating uniform width; (b) inclination of the penholder to one side of the drawing direction results in the opposite side of the line being irregular; (c) drawing the start and finish of a line too slowly; (d) drawing too fast results in variation of width; (e) poor ink flow leads to the creation of irregular width linework; (f) too much ink at the tip of the point results in 'blobs' being produced.
② Line connection: (a) overlapping one line with another tends to produce a bulge at the point of contact. To avoid this, drawing speed should be increased at this junction; (b) the exact connection of lines produces a much better result; (c) if two lines do not join precisely, the gap between them can be filled by the repeated application of thinner lines.

4.2.10.6   The Pelikan 'graphos'
Shown in Fig. 4.104.

4.2.10.7   Using a technical pen/tubular pen, with an ink cartridge, in a compass
Shown in Fig. 4.105.

4.2.10.8   Using a technical pen and a straight-edge to produce linework of constant width
Shown in Fig. 4.106.

4.2.10.9   Commercially available drawing ink
① Filler cap.

4.2.10.10   Ink-pot for use on a drawing desk
① Ink reservoir; ② water reservoir and sponge.

4.2.10.11   Commercially available ink dropper bottle (plastic)
Shown in Fig. 4.109.

4.2.10.12   Filling a cartridge-type reservoir with ink
Shown in Fig. 4.110.

4.2.10.13   Technical pen incorporating a piston-type filling mechanism
Shown in Fig. 4.111.

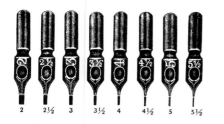

Fig. 4.99

Fig. 4.100

Fig. 4.101

Fig. 4.102

Fig. 4.103

Fig. 4.104

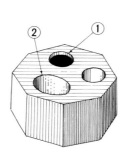

Fig. 4.105

Fig. 4.106

Fig. 4.107

Fig. 4.108

Fig. 4.109

Fig. 4.110

Fig. 4.111

Table 4.1

The surfaces of the polyvinyl and polyester plastics have been prepared during manufacture to enable the creation of high quality linework. For example, polyester has either a matt or polished surface.

Base Material	Dimensional Stability	Trans-parency	Flexibility	Danger of breakage or tearing	Inflamma-bility	Foldability	Ease of ink erasing	Other properties
Glass plates with laquered surface	excellent	high	none	easily broken	none	none	good with scraper	bulky
Metal plates laminated with drawing paper	very good	opaque	poor	none	none	none	poor with rubber	
Drawing paper	poor	opaque	very good	tears easily	fair	fair	poor with rubber	
Tracing paper	poor	translucent	very good	tears easily	fair	fair	poor with rubber	crumples when wet
Polyvinyl plastics*	good	highly translucent or opaque	good	splinters	high	low	good with scraper	static elec-tricity melts at 55°C
Polyester plastics*	good	highly translucent or opaque	very good	none	high	high	good with scraper or wet rudder	unreceptive to dyes, melts at 150°C

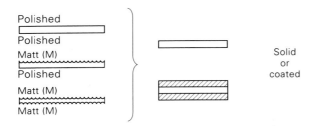

(a)

(b)

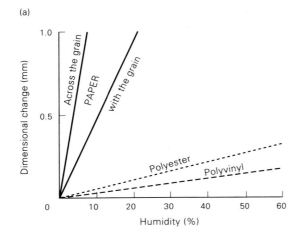

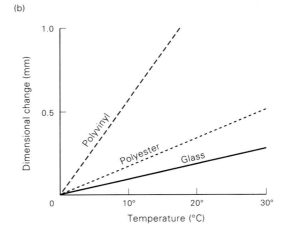

Fig. 4.112(a) Dimensional changes in various materials due to changes in humidities

Fig. 4.112(b) Dimensional changes in various materials due to changes in temperature

4.2.11 Drawing base materials

Materials most commonly employed as bases during the generation of maps are shown in Table 4.1. Information with regard to their dimensional stability is graphed in Figs 4.112(a) and (b). Currently, the superior qualities of polyester film make it the material most widely used as a base during map preparation.

The polished surface of normal polyester film is rendered matt by physical or chemical graining in order to improve its receptivity to pencil, ink or typing; alternatively, it can be pre-lacquered to weaken its resistance to scribecoat engraving, the peeling away of a masking membrane, or the removal of a light-sensitive emulsion.

There are many varieties of polyester drawing film which are pre-processed to satisfy specific user requirements. Typical examples of commercially available products are provided by Kimoto's AK films and the K. & E. Stabilene range.

4.3 Drawing in ink

4.3.1 Constructing simple symbols

The basic methods relating to the drawing of symbols, using pen and ink, are explained in this section. The icons described have been selected from those commonly employed in topographic mapping. Japanese practices have been illustrated to provide particular examples.

4.3.1.1 The creation of small, individual symbols
① Horizontal lines are drawn, from left to right, using a pen and either a glass straight-edge or a set-square. This process is further described in paragraphs 4.2.4.4 and 4.2.4.5. Subsequently, the vertical line is drawn freehand, from the top down. ② All horizontal lines are drawn sequentially, starting from the top and working downwards. Vertical lines are then produced, successively, from the left to the right. ③ Symmetry and

balance with respect to the arrangement of the lines is estimated by eye.

4.3.1.2 Drawing small point symbols with standard shapes
① A triangulation point: first position a dot; next, using a ruler, construct a sloping line to the left of this; then, either freehand or with the aid of a straight-edge and by working from top to bottom, produce a second line to the right of the dot and with the same angle of inclination as that on the left. The base can then be drawn in. Care must be taken to ensure that the dot is located at the centre of gravity of the resultant triangle. ② A square: horizontal lines are constructed using a straight-edge and vertical lines are drawn freehand. The two sets of lines must meet at the corners as right angles. ③ Small circles: these are drawn in a clockwise direction as is explained in paragraphs 4.2.8.5 to 4.2.8.8. The innermost

element of concentric circles is produced first, and all lines drawn later must be precisely generated using the same centre and horizontal starting points.

4.3.1.3　The sequence of operations involved in drawing small symbols composed of lines
① Those consisting essentially of oblique lines and a horizontal tick (e.g. a coniferous tree symbol). Inclined elements are drawn freehand on both the left and right sides, but the horizontal tick is produced with a straight-edge. ② Devices comprised of both oblique and vertical lines together with a horizontal tick (e.g. the symbol for a mulberry tree). The first two of the three line types are drawn freehand, and the final one with a straight-edge. ③ Symbols consisting of both horizontal and vertical lines (e.g. rice paddy). Short horizontal elements are drawn with a straight-edge, and verticals freehand. This order of operations helps to keep the ends of vertical lines level. ④ The combination of vertical and oblique lines with a horizontal tick (e.g. the icon for bamboo). The horizontal tick is drawn with a straight-edge, but all other work is produced freehand. ⑤ Series of parallel vertical lines (e.g. a grass symbol). All lines are produced freehand, with the first completed being the longest one in the centre. The next drawn lies to the left of this, and the final one to the right.

4.3.1.4　Small circular symbols
① Small circle with a horizontal tick (e.g. the symbol for a broadleaf tree). Construct the circle using a drop compass, and the tick with a straight-edge. ② Small circle with a vertical tick (e.g. an orchard symbol). Construct the circle using a drop compass, and the tick freehand.

4.3.1.5　Dot symbols
① A multi-dot symbol (e.g. a symbol representing a tea bush). Draw three dots, all of the same size, in the order shown. ② Another multi-dot symbol arrangement (e.g. a lawn symbol). Draw three dots, all of different sizes, in a horizontal straight line by working from left to right.

4.3.1.6　The arrangement of regularly spaced symbols
The positioning of symbols in a standardised way, for example to represent the distribution of cultivated crops, is accomplished by inserting a sheet of positive film incorporating a regular grid under the working drawing. An appropriate interval between symbols is selected, and each is then positioned with reference to the underlying grid.

4.3.1.7　Construction of a dotted line
① Individual dots should exhibit uniform size and spacing, and be drawn by working from left to

right; ② irregularly sized dots and spaces, or failure to maintain their alignment, results in an apparent loss of linear continuation; ③ dots are drawn by moving the tip of the pen in a clockwise direction.

4.3.1.8　Construction of a pecked line
① Pecked lines are produced by working from left to right; ② draw solid elements of equal length, and leave the same amount of space between each of them. The alignment of the ends of each of the pecks must follow the overall line direction.

4.3.1.9　Stages in the construction of a double-line symbol
① Draw parallel lines with a double ruling pen (road pen); ② insert equally spaced ticks between the lines; 3 opaque alternate segments to produce a railway line symbol.

4.3.1.10　Stages in the construction of decorative pecked lines
① Draw a pecked line; ② position ticks at both ends of the solid elements; ③ add a dot in the centre of each of the spaces between pecks.

4.3.1.11　Stages in the construction of linework consisting of three differently sized dots
① Draw a line composed of the largest size of equally spaced dots to form the outer edge; ② place a line of medium sized, equally spaced dots below this, with the position of each solid element coinciding with the centre of the space between dots generated in ①; ③ produce a line consisting of the smallest size of equally spaced dots to form the inner edge of the composite symbol. Each solid element must coincide with the centre of the space between the dots comprising ②.

4.3.2 Drawing symbols in combination

When required to generate topographic mapping, the drawing methods used to demonstrate the crossing or connection of linear symbols; the duplication of linework when related to a specific place; and the positioning of point symbols relative to particular locations; are all based on established cartographic principles. Thus, mapped detail must appear precise, reliable and not open to doubt or misinterpretation. In addition, it should be clearly visible and easily readable. Included symbolisation must be well designed and drawn in order to facilitate communication and to enable the reader's immediate awareness of the exact position, shape, size and characteristics of features, and also the spatial relationships existing between them, etc. The essential points influencing these factors are illustrated by examples described in

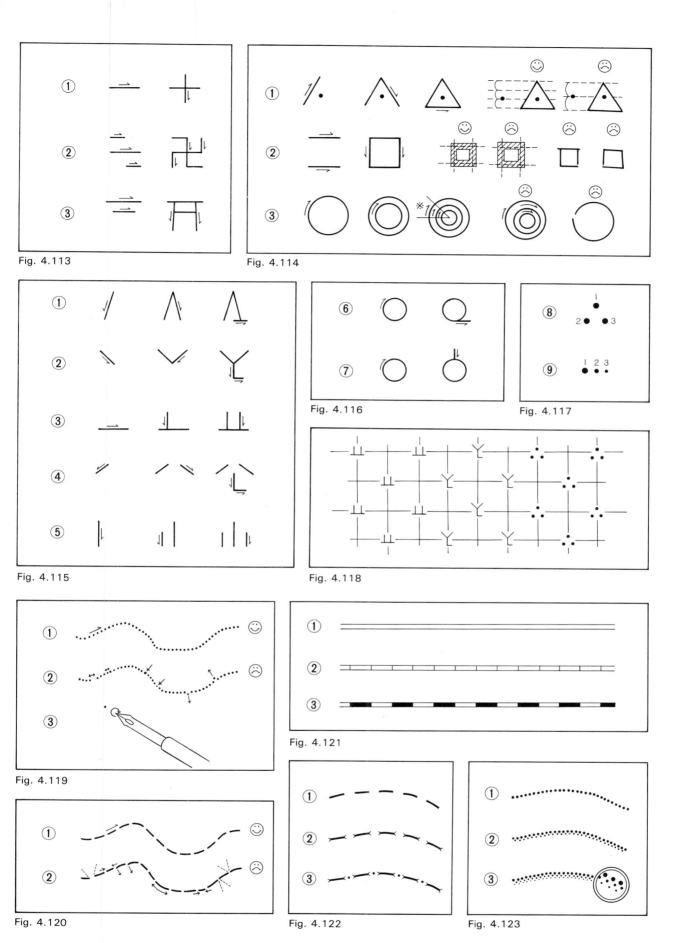

Fig. 4.113

Fig. 4.114

Fig. 4.115

Fig. 4.116

Fig. 4.117

Fig. 4.118

Fig. 4.119

Fig. 4.121

Fig. 4.120

Fig. 4.122

Fig. 4.123

127

the following paragraphs. Some of the employed techniques may not correspond exactly with internationally accepted standards. This results from differences in the purposes for which maps are required — national traditions, etc.

4.3.2.1 Railway symbolisation

Some 'local' modification to the spacing between ticks, or the black and white segments, drawn to represent railway lines is sometimes necessary to create an appropriate impression which is both clear and legible. Typical examples are included here. The techniques cited are also relevant to all discontinuous line symbols as is further explained in paragraphs 4.3.2.2 to 4.3.2.4.

(1) Ticks positioned along parallel straight lines must be regularly spaced, but their location at junctions should be avoided; (2) an irregular arrangement of ticks creates an unnecessarily complicated visual impression and affects readability; (3) alternating black and white segments should be regularly spaced, but the junction of two lines must always occur within a white element.

4.3.2.2 Ways of connecting different types of line symbols

Methods relating to road symbols are described below. Some techniques require local adjustment to be made with respect to the position of intervals between pecks.

(1) The junctions between road symbols of different types or widths must be carefully drawn, and the centre lines should coincide; (2) when using a double-line road symbol any included pecks must conform with the rules relating to single pecked lines. This results in an unambiguous impression of the continuity of a road network; (3) solid elements within single pecked lines must be positioned at junctions or corners for the reason stated in (2) above. Alternatively, for example on maps of mountainous areas, it may be appropriate to locate a gap at a junction or crossing point, so allowing the inclusion of other identifying detail.

4.3.2.3 Various examples of crossing line symbols

(1) A railway passing over a road at a level crossing; (2) a road bridge; (3) a road bridge over a railway; (4) a railway bridge over a river. An alternate method of showing this involves using the same symbol as that for a road bridge; (5) a bridge symbol may be omitted when a stream is small; (6) it is preferable to locate a black segment at the entrance to or exit from a tunnel; (7) gaps between pecks should be located at points where two symbols cross; (8) the symbol depicting a feature such as an electricity transmission line is normally omitted where it crosses another icon.

The following paragraphs provide further elaboration of methods described in (2) of 4.3.2.2.

4.3.2.4 Constructing pecked double-line symbols

Shown in Fig. 4.127.

4.3.2.5 Drawing double-line roads in association with other symbols

There are two methods for the generation of hachures depicting a road embankment. In the first, the tops of the hachures are joined to the edge of the carriageway, but in the other a line connecting the tops is positioned slightly away from the road. The former is best for monochrome mapping in that it saves space and depicts the position of the feature precisely. The latter is used in multi-colour production and allows easy adjustment during printing and clear identification of the hachures.

In built-up areas a single-line road may be replaced by a narrow, double-line street symbol.

4.3.2.6 The location of individual point symbols

(1) An attribute symbol relating to a building must be positioned either inside or just above it. Failure to do this makes it more difficult for the map user to identify attributes. (2) Point symbols representing individual objects must be located in their correct position. This can be related to the bottom left corner or centre of the base in instances where an object is obliquely represented; or the geometric centre if they are viewed as if from directly above.

4.3.2.7 Contouring

(1) Using a contour pen, draw lines from left to right and as a series of strokes. These are subsequently linked at the centres of the sharpest corners, such as those occurring along the course of a valley or a narrow ridge. The attempted joining of strokes within areas of shallow curvature often results in a very evident failure to achieve a clean connection. (2) Draw lines, working from the highest to lowest in all areas of high relief. Strokes should be made from one narrow ridge or valley line to the next such adjacent feature.

4.3.2.8 The relationship between contours and road symbols

Link all contour lines with double-line roads or streams at as near to a right angle as possible. Strange results occur if this practice is not followed.

4.3.2.9 Boundaries in association with linear features

(1) Boundaries running along the centres of rivers and streams, or in areas of open water, are normally indicated by drawing segments of a specific length at all points where they change direction; (2) intermediate areas between segments, or along the crest of a narrow ridge, are usually omitted.

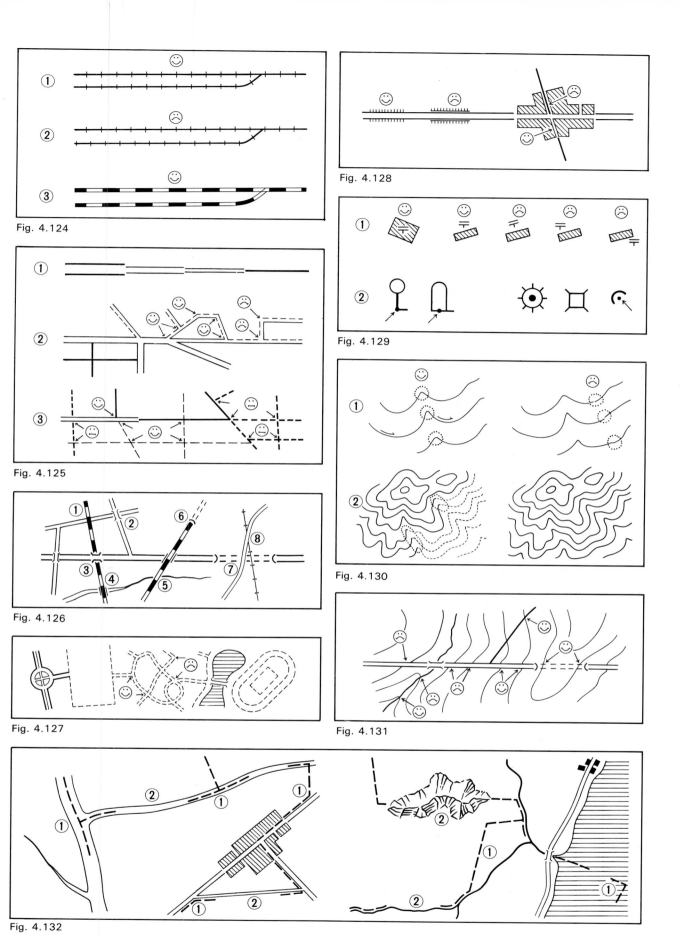

Fig. 4.124

Fig. 4.125

Fig. 4.126

Fig. 4.127

Fig. 4.128

Fig. 4.129

Fig. 4.130

Fig. 4.131

Fig. 4.132

4.3.3 Hachures

This technique for the representation of terrestrial relief on topographic maps was originally developed in the mid 17th-Century, and was more correctly termed 'vertical hatching'. It was the dominant method used to depict landforms on maps produced in the 19th Century, because the employed copper plates from which they were printed provided an eminently suitable medium for the engraving of fine linework. There are two methods of hachuring — vertical and horizontal. They are still used, to some extent, in modern topographic mapping for the depiction of small hills, embankments and cuttings.

4.3.3.1 Constructing hachures

① Each of the arrows indicates the direction of slope. The degree of steepness, from place to place, can be estimated with reference to the distances between contour lines. ② Evenly spaced hachures are drawn as short lines in accordance with each of the lines of maximum slope. The position of each row of hachures is slightly offset from that of the one above it, and there is a narrow gap between each of the belts. The width of the individual lines is decided with reference to one of the methods illustrated in ③ and ④. ③ Vertical illumination: variations in the thicknesses of lines are completely dependent on the angle of slope. This means that the direction from which it is lit must be at right angles to the surface. The width of hachures is decided with the aid of a scale which is usually in the form of a personally produced graphic table. It consists of several categories of line widths which have been determined with respect to angles of slope. ④ Oblique illumination: the direction of lighting is normally at 45° to the upper left-hand corner of the sheet, and at an angle of 45° to the surface. Hachure thickness is decided with reference to the amount of light reaching each part of the landscape, this being dependent on the angle of slope and its position relative to the direction of the source of illumination. This technique provides a more three-dimensional effect than vertical lighting, and is the one normally used in topographic mapping.

4.3.3.2 The spacing of hachures

Hachure width is decided by maximum angles of slope; typical examples are shown in Fig. 4.134: ① 30°; ② 20°; ③ 10°; ④ 5°.

A virtually constant interval is maintained between hachures. In the left-hand illustration it is 20 lines per 10 mm, whilst on the right this increases to 40 lines per 10 mm. Both graphics have been subjected to 10 times photographic enlargement.

4.3.3.3 Difference resulting from the use of each method of illumination

The visual differences created by applying ① vertical illumination and ② oblique illumination are demonstrated here.

4.3.3.4 The relationship between hachure width and slope angle

① Lehmann's system. In 1799 an Austrian army officer named Johann Georg Lehmann developed a formula relating hachure width to maximum angles of slope. The underlying principle of his system is 'the steeper the darker'. The ratio of black to white, representing the degree of slope, is varied from 0.0 to 1.0 in ten equal steps, as is shown here. ② A modification of Lehmann's system. This was introduced by a Russian cartographer, P. Bolotov, in a publication entitled *Osnovy topografii i kartografii* which appeared in 1959. It is applied to the illustration of a landscape where gentle slopes predominate. ③ A second modification to Lehmann's system. This was also introduced by a Russian cartographer, called Glabnog, in the text cited above. It too relates specifically to the depiction of gently sloping relief. This technique can be applied to the design of an actual hachuring scale, and also a shading gauge as described in subsection 4.3.5.

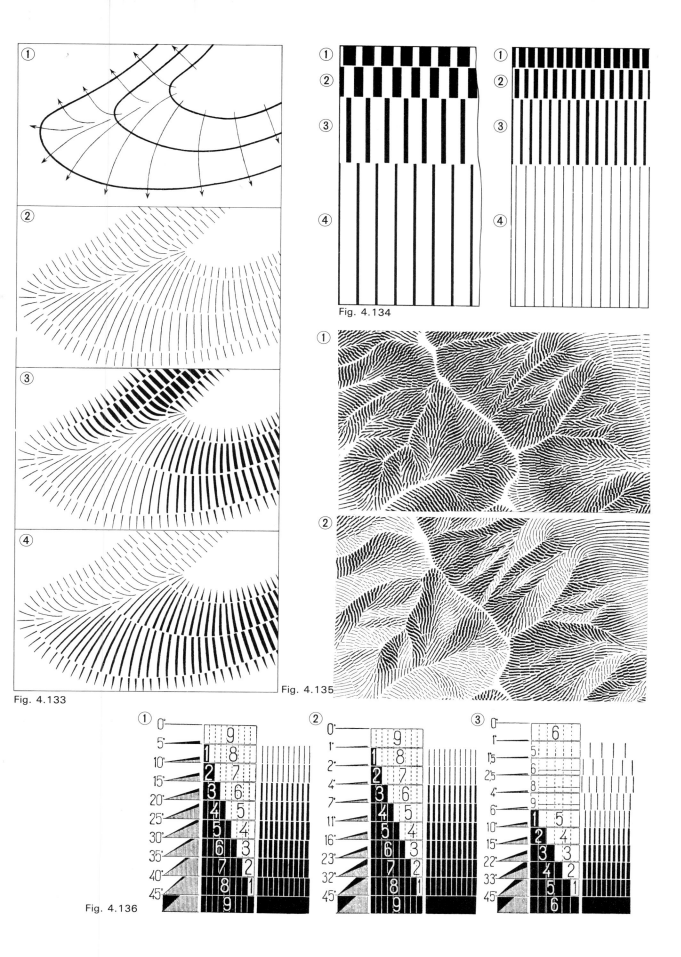

Fig. 4.133

Fig. 4.134

Fig. 4.135

Fig. 4.136

4.3.4 The application of hachuring

4.3.4.1 *Representing steep slopes*

The length, width and interval between hachures used to depict small but steep slopes, on maps at varying scales, has been standardised by E. Imhof (1965) as shown in Table 4.2.

Table 4.2

Scale	Length (min.) (mm)	Width (mm)	Interval (mm)
1 : 2 000 and larger	0.5	0.5−0.3	1.0−0.6
1 : 5 000	0.5	0.4−0.2	0.7−0.5
1 : 10 000	0.4	0.3−0.2	0.5−0.4
1 : 20 000 and smaller	0.3	0.2−0.1	0.3−0.2

The illustrations numbered ① − ⑱ consist of good and poor examples of the use of hachuring. ① Provides a real impression of a slope and clearly indicates its top and bottom, whereas ② to ⑤ do not. However, ④ and ⑤ are easy to draw and could well be used on large-scale maps where it is less necessary to give as much emphasis to landscape character; ⑥ creates a better three-dimensional impression than ⑦; ⑧ suggests a concave slope rather more than either ⑨ or ⑩; ⑪ allows the easy perception of contours used in association with the hachures, whilst ⑫ is poor and rather less clear; ⑬ is an acceptable depiction of the relationship between a contour and a slope symbol, but ⑭ is not; ⑮ provides a clear illustration of a small slope in association with a contour, but ⑯ is not so successful as the two symbols have merged to produce a sort of smear; ⑰ is a good example of small slope portrayal as compared with ⑱, for the same reason as ⑮ and ⑯.

4.3.4.2 *The representation of rocks and cliffs*

Rock fields and cliffs occurring in steep, mountainous areas are illustrated by sketches. These consist of a combination of vertical and horizontal lines which give a pictorial impression of the feature which they represent. Some good and poor examples of how this may be achieved are shown in the examples. ① This is good in comparison with ④ or ⑦. ② This is good in comparison with ⑤ or ⑧. ③ This is good in comparison with ⑥. ⑨ This is good in comparison with ⑩ or ⑪. ⑫ This is good in comparison with ⑬ or ⑭. 15 This is good in comparison with ⑯ or ⑰.

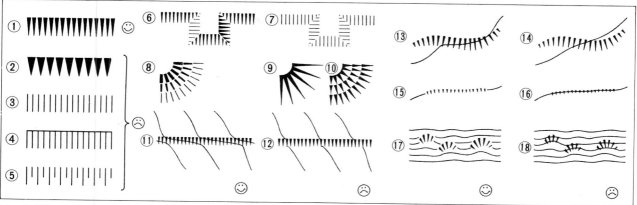

Fig. 4.137 Representing steep slopes

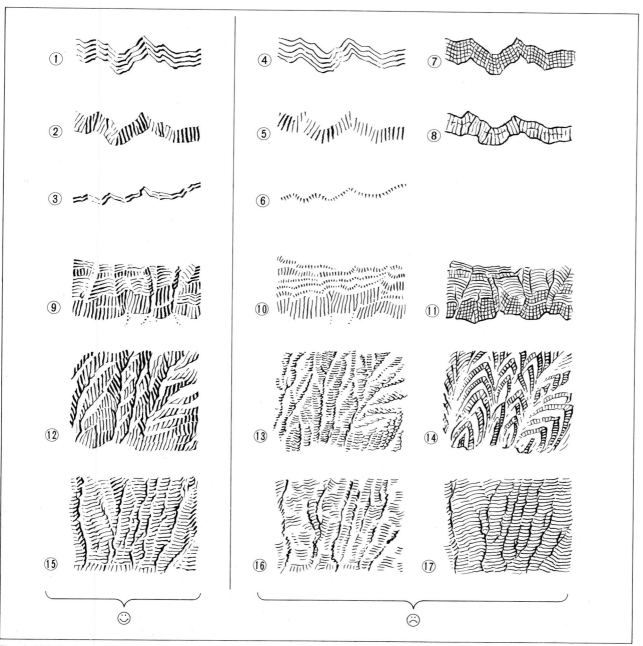

Fig. 4.138 Rock and cliff depiction

133

4.3.5 Principles of hill-shading

This technique is employed in mapping to represent surface relief by the creation of a three-dimensional visual impression. It is used as an alternative to hachures, and the inherent tonal variations are dependent on the orientation of each element of the landscape with respect to the position of a light source. Further, the darkness of shading is related to the angle of slope, i.e. the density of contours, as is illustrated in Fig. 4.139 which is based on vertical illumination. In the case of a landscape receiving light at an oblique angle, the perceived darkness of each relief feature varies according to the angle it subtends with the direction of illumination.

In Fig. 4.139 ① five classes of slope, ranging from steep to gentle, and a horizontal plane are depicted by contours. The spacing between these increases from left to right. ② Demonstrates the results of lithographic printing, as compared to ③ which consists of original drawing. Each of the steps relates to the actual degree of slope shown in ①. The contrast between the printed tones of ② is weaker than that of those drawn in ③. The horizontal plane is represented by the lightest tones in both ② and ③. All densities of shading are printed, by lithography, on white paper using a prescribed screen of known density. In consequence, the shading constituting the original drawing is not faithfully reproduced, and this causes problems with respect to methods used for both constructing the graphic and its reproduction. A possible solution can be provided by preparing three successive originals. The first of these incorporates conventional tonal contrasts over the complete area mapped; the second serves to emphasise contrasts in specific areas, and the final one acts as a mask and includes windows for the highlighted areas.

With regard to drawing techniques, the appropriate darkness or lightness of shading prepared may, as is demonstrated in Fig. 4.140, require the modification of a theoretically correct representation to improve its resultant appearance. In the upper half of the figure the facets facing the direction of illumination (a–b) are shown in the same tone and appear as one continuous, uniform element. The lower example has been improved by the partial modification of direction of illumination in order to assist the distinction of individual facets. Tones on horizontal surfaces have also been lightened to improve the clarity and visibility of the complete figure.

Differences in the three-dimensional appearance, resulting from the application of different methods of shading and hachuring, are shown in Fig. 4.141(a)–(f). Figure 4.141(a) is the side-view of an object; Fig. 4.141(b) its theoretical shaded appearance when subjected to illumination from the north-west corner; Fig. 4.141(c) is the equivalent version illustrated by hachuring. In Fig. 4.14(d) a conventional method of graphic depiction has been used to portray the object using an oblique light source, but tone has been omitted from the horizontal surface to help make the representation clearer. Figure 4.141(e) and (f) relate to the same object subjected to a vertical source of illumination. No three-dimensional impression is evident because all of the slopes exhibit the same degree of inclination.

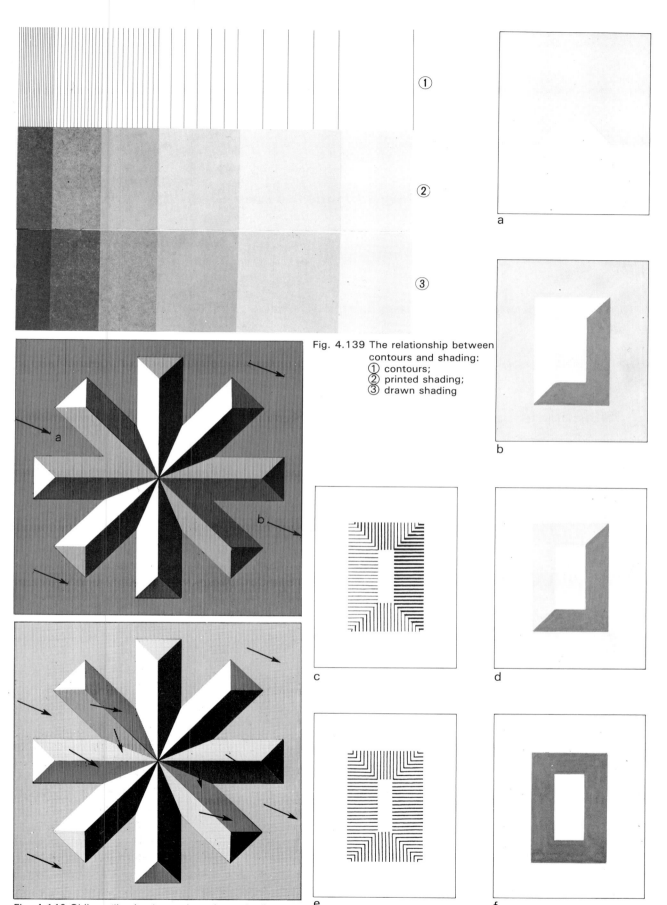

Fig. 4.139 The relationship between
contours and shading:
① contours;
② printed shading;
③ drawn shading

Fig. 4.140 Oblique illumination and resultant shading.
Upper: theoretical; lower: modification of theoretical
impression to emphasise three-dimensional appearance
and improve map clarity

Fig. 4.141(a)−(f) Differences in three-dimensional
appearance generated by applying various shading and
hachuring methods

135

4.3.6 Differences resulting from variations in methods of illumination

Employed shading techniques can be combined with different methods of lighting in order to create an appropriate impression of relief features represented at a particular map scale.

Figures 4.143 and 4.144 are shadings produced to be used in association with the large-scale (1 : 25 000) contour map shown as Fig. 4.142. Figure 4.143 has been drawn to highlight elements facing the direction of illumination and the flatter areas, and the tops of hills, appear in a medium tone. This 'theoretical method' involves imagined oblique lighting (emanating from the north-west) shining on the mapped area. In Fig. 4.144 the shading has been generated by using a combination of both vertical and oblique illumination. Thus the flatter and more nearly horizontal areas are not toned, but rather highlighted, and all slopes are shaded as if subject to standard, oblique lighting. This technique, involving a combination of conventional/vertical and theoretical/oblique illumination, is often used to produce a clearer impression than would the latter if it were to be applied in isolation. The normal role of shading is to provide a three-dimensional impression of relief which acts as a background to other mapped detail. Consequently, the tones employed should not interfere with the potential visibility of other items, unless the map's main purpose is to communicate information about the character of a landscape. In this case the shading drawn must be closely based on the results of oblique illumination.

Figures 4.146 and 4.147 are shaded representations relating to a small-scale contoured map of an area at 1:1 000 000, which appears as Fig. 4.145. As with Figs. 4.143 and 4.144, Fig. 4.146 is based on theoretical/oblique illumination, and Fig. 4.147 also uses the conventional/ vertical method. On large-scale maps, slope features are emphasised by variations in tones, whereas at small scales global relief types such as mountain ranges, peaks, rugged areas, etc., are stressed, rather than less physically significant items. In the latter case, additional tonal contrasts must be introduced, with darker shadings being used for high mountains and lighter and softer ones for lower lands. By doing this a three-dimensional impression can be created for the whole of the area mapped.

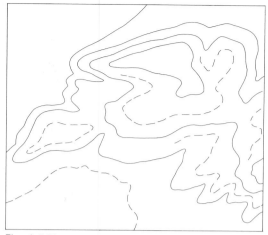

Fig. 4.142

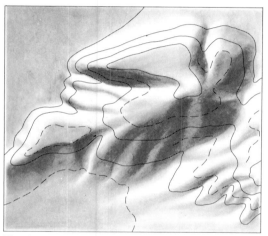

Fig. 4.143

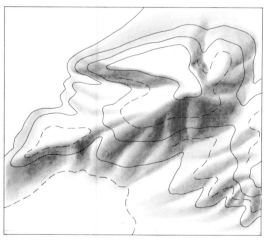

Fig. 4.144

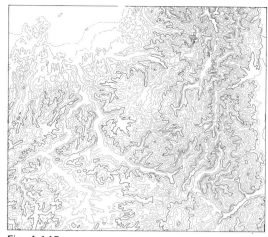

Fig. 4.145

Fig. 4.146

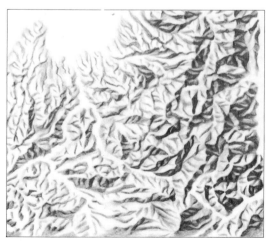

Fig. 4.147

137

4.3.7 Shading to illustrate various types of topography

Good and poor examples of shading used to illustrate different types of topography are provided in Figs 4.148 to 4.150. All of these have been drawn based on the imagined use of an oblique light source shining from the north-west. Figure 4.148 represents a high, mountainous area which includes a number of narrow ridges. The illustration on the left gives an appropriate impression of the ridges, but that on the right does not. An upland area subject to down-cutting erosion is shown in Fig. 4.149, and Fig. 4.150 features sketetal hills. In each instance ① depicts the area by a combination of contours and shading; ② to ④ picture the same area as ①, but are shaded at the reduced scale of 1/2.4. Characteristic features of the topography are correctly stressed in ④ (in both cases), but ② and ③ suggest a different type of landscape.

4.3.8 The production of hill-shading

Shading provides a typical example of the generation of continuous-tone imagery. When completed, it must, of necessity, be converted to discontinuous/discrete material to enable its reproduction by lithographic printing. This modification is normally undertaken photographically, and involves the use of a half-tone screen. The continuous-tone original can be produced by using a pencil, a brush to apply well-mixed paint, or an airbrush to deposit paint or ink. The employed base material can be either a translucent white plastic sheet or a matt-surfaced bromide paper. Drawing involves the accurate positioning of the desired feature, and its depiction by the application of shading of an appropriate tone. In order to control density, a graphic key should be prepared which is similar in character to the grey scale shown in Fig. 4.139

①.

The most common method of producing hill-shading requires the use of different grades of pencil and grey paper stubs. Drawing takes place as a result of controlled pressure and repeated pencil use. Light tones are produced by the application of a medium grade lead with gentle pressure, and dark shading requires more pressure and a softer lead. A grey paper stub is used to smooth the drawing as is shown in Fig. 4.151. The depiction of intricate detail on small-scale maps is most easily achieved by pencil drawing, but this may be supplemented by pen and ink or brush-work if necessary.

When using a brush, the resultant tone depends on the proportion of pigment to water, or the mixture of pigment and white in the form of a gouache. The dilution of water-colour paints changes their transparency and so allows differ-

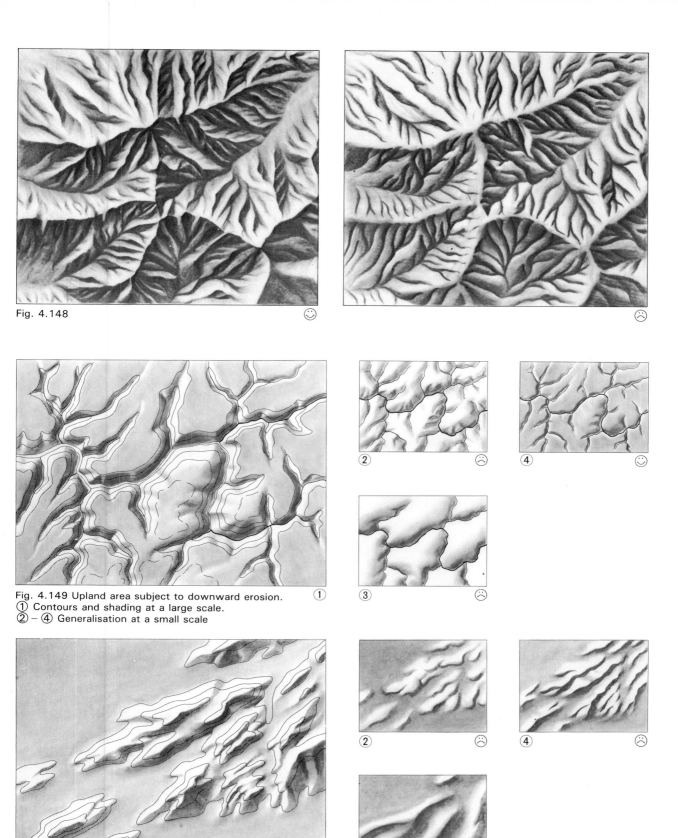

Fig. 4.148

Fig. 4.149 Upland area subject to downward erosion.
① Contours and shading at a large scale.
② – ④ Generalisation at a small scale

Fig. 4.150 Skeletal hills

139

ent tonal values to be generated.

It is difficult to achieve smooth tonal gradation using only a brush, and so an airbrush is often applied. This sprays fine particles of pigment, held in a solution, over the surface, and the density of the paint and duration of spraying control the build-up of different tones. In the most common instruments a small quantity of paint or ink is contained in a reservoir and is drawn, under pressure, into a fine tube from where it is expelled through a nozzle. The area covered by the jet depends partly on its distance from the image, but also on finger control of the airbrush. Shading for medium and large-scale maps without small or intricate areas of relief is most easily produced by this method. The instrument illustrated in Figs 4.152 and 4.153 allows considerable control to be exercised over the fineness of the jet, and consequently can be employed to produce linework as narrow as 0.1 mm. Thus it could be used in preparing the very detailed shading typically appearing on small-scale mapping.

An airbrush is also available for the manual production of a vignette as is shown in Figs 4.153 and 4.154. The term 'vignetting' relates to the creation of a band of colour to emphasise a line bounding an area. Its tone gradually decreases in strength away from a line, such as a coast (Fig. 4.154) or the boundary of a wooded area.

4.3.8.1 The relief contour or orthographic relief method

Professor Kitiro Tanaka developed four methods for the production of relief shading. These are the orthographic, relief/illuminated contours, relief hachuring and relief hachuring combined with hypsometric tints. He also established the mathematical theories on which the techniques are based. Information on the relief contour method was first published in the Japanese *Geographical Review*, 15, 9–10 (1939), and later in the *Geographical Review*, 40, 444–456 (1950).

The relief/illuminated contour method is now well known, but an example appears here as Fig. 4.155. It gives an impression of a stepped relief model, and the principles relating to its generation are illustrated in Fig. 4.156. The top drawing is a side view of a hemisphere positioned on a plane surface. ① This is the horizontal plane forming the base; ② shows lines representing contours; and ③ is the light source which is positioned at an angle of 45° to the base—as is normal when oblique illumination is used in more traditional methods.

The second diagram provides a plan view as it would appear in the illuminated contour method: ④ is a contour layer, the normal position of which (relative to the source of light) is denoted by the horizontal angle θ; ⑤ is the base which is shaded with a medium tone; ⑥ indicates contour layers

facing the light; and ⑦ those facing away from it. In the drawings at the bottom, that on the left illustrates the ratio between the thickness of drawn lines (t) and the horizontal interval between contours (d); whilst on the right can be seen the relationship between the width of the nib of a pen used for drawing (t_o), t and θ. In these graphics ③ shows the direction from which illumination is received; ④ is the normal position of a contour line of specified width (Δl); ⑤ is the horizontal base, shaded using a medium tone; ⑥ is a contour of known width, t; and ⑧ represents a broad-nibbed drawing pen with a width of t_o. The relevant formula for contour construction is expressed as $t = t_o \cos \theta$.

The actual method of drawing involves following a contour guide trace with the pen, and keeping the broad-nib parallel to the direction from which illumination originates ③. The value of t_o is obtained by using a formula which is as follows: B is the oblique angle brightness value ratio, which is the cosine of an included angle between the normal position of a contour and the light. The latter illuminates the base at 45° and the slope angle between contours at γ, then:

$$B = (\sin \gamma \times \cos \theta + \cos \gamma)/\sqrt{2}.$$

Further, the precise brightness value (Bm) relating to a contour layer with a vertical interval of d_o, at the map scale, can be expressed by means of a formula in which: t_o = the maximum width of the contours drawn; k and K = the constant ratios; b = the mean brightness of a layer; bm = the maximum mean brightness of a layer with its normal orientation parallel to the light source ($\theta = 0$ and $\gamma = 45°$); p = the brightness of the layer unit; and p_o = the tone of the base. Thus:

Bm = b/bm = K [$p_o + k (p - p_o) \cos \theta \times \tan \gamma$] / K [$p_o + k (p - p_o)$] = (1 + $\cos \theta \times \tan \gamma$)/2. Here $t = p_o/(p - p_o)$ (coefficient of contours). $t_o = kd_o$ and $t = kd_o \cos \theta$. In terms of B and Bm, the difference ratio E is expressed as follows:

$$E = (B - Bm) / B = 1 - (\sec \gamma) / \sqrt{2}.$$

The value of E is approximately 29% when $\gamma = 0$, and is smaller when γ is larger. E is also independent of differences of θ. The brightness of the base must be maintained at 50% when it is shaded in grey or another colour such as greenish yellow or yellowish red. In addition, the value of k can be changed from the theoretical one. If the actual k is made three times larger than the theoretical value, the resultant three-dimensional impression will be three times stronger.

The final graphic is generated by combining two drawings. One of these represents relief contours on the side facing the source of illumination, whilst the other displays those facing away from it. The original drawings are superimposed at the lithographic printing stage.

Fig. 4.151 Smoothing pencil drawing with a grey paper stub

Fig. 4.152 Shading with an airbrush

Fig. 4.153 Vignetting with an airbrush

Fig. 4.154 Vignetted coastline

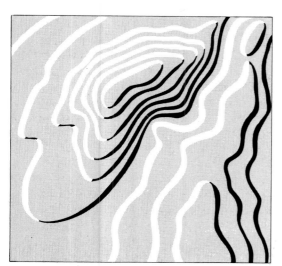
Fig. 4.155 Example of an orthographical relief map

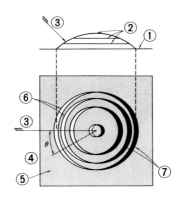

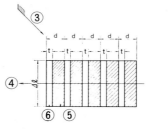

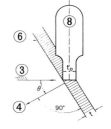
Fig. 4.156 Principles of the Kitiro Tanaka method (orthographic relief method)

4.4 Scribing

This is a cartographic drawing method which culminates in the production of original negative films which are used in the manufacture of lithographic printing plates. It has largely taken over from traditional pen and ink drawing and its subsequent photography. Scribing on glass, and later plastic sheets, became widespread in the late 1940s and 1950s, and was a natural development from wet-plate photography. It allowed the touching-up and correction of images as in the colour separation process. Japan, and many European countries, changed from pen and ink working to scribing during this period.

Perhaps the best examples of the use of glass are provided by the most excellent Swiss topographic maps which include contours printed in three different colours. Black was used to show relief in rocky areas, blue for permanent glaciers, and brown elsewhere. Resultant linework is very sharp and the registration between colours is carefully maintained, particularly in very mountainous areas where contours are dense and perhaps as little as 0.2 mm apart. This degree of accuracy could only be achieved by scribing on glass and creating separate drawings for each of the colours. In order to do this, guide copies of the original compilation were printed down on the glass plates required to produce each of the three coloured contours. After scribing all of the detail to be shown on a specific plate, the engraved glass was contact-printed in combination with the information intended to appear on the next plate in a different colour. The second sheet was then scribed with the new linework being precisely positioned in relation to that already completed. The procedure was repeated for all other required contour detail. The process is still employed today, using polyester plastic scribecoat, in cases where different colour linework separations have necessarily to be printed in very close proximity.

Initially, polyvinyl chloride was used as the base material for plastic scribecoat. The technique was introduced by S. Sachs of the US Coast and Geodetic Survey, and was widely adopted by other American governmental cartographic agencies. Later, polyvinyl was replaced as the standard base by polyester film which was readily available commercially and more dimensionally stable. Pre-punch registration systems were also developed and virtually overcame the problems relating to the accurate assembly of fine, scribed detail on different plates.

Linework scribing is used in combination with masking to generate area patterns, and with stick-up or dry transfer lettering. It is employed, in conjunction with a pre-punch register system, to produce a set of standardised original drawings. A complete set of finished films can be used not only in photolithography and the printing of large numbers of copies of multi-coloured maps, but today is also utilised in photo-contacting and/or electrostatic copying to print off small quantities of mono- or multi-colour documents.

4.4.1 Scribing points

4.4.1.1 Types of scribing points
Two varieties of scribing points can be distinguished in terms of their shape. The first is a conical-tipped needle which is used to produce fine lines, and the other, which exhibits a chisel point, is employed in the construction of thick, bold, or double and triple lines. The relationship between the two types, and the relative widths of lines which they are used to produce, is demonstrated in Fig. 4.157 ('d' indicates the width of a scribed line).

① Conical-tipped needle for a fine line: d = 0.15 mm or less; the outline of the tip is slightly rounded; ② conical-tipped needle for a thick line: d = 0.15−0.20 mm; ③ chisel point for a bold line front view): d = 0.20 mm and wider; ④ chisel point (side view).

All needle points are made to a specified thickness and, when correctly used in conjunction with either a pen-type or a fixed scriber, will produce linework of the defined width. Chisel points exhibit similar characteristics, but are employed to generate bolder and double or triple lines of a known thickness, whilst held in a swivel scriber.

Commercially manufactured scribing points (see Fig. 4.173) are now in general use. However, in some parts of the world their acquisition may be difficult; they may be prohibitively expensive; certain sizes or varieties of points may not be obtainable; or, where they are available, have become worn down after extensive use. In consequence, it may become necessary to shape or sharpen scribers in the drawing office.

4.4.1.2 Shaping the point of a needle
The sequence of operations is illustrated by numbers ① − ⑦: ① a sewing needle, steel gramophone needle or drill rod is clamped in a chuck; ② the tip is heated with a spirit lamp to temper it; ③ a sharpening stone is used to grind the tip; ④ the point is re-heated; ⑤ mechanical oil is used to temper it; ⑥ the point is tested on a sheet of scribecoat and the width of the line is checked using a microscope with 20× magnification; ⑦ the needle is fixed in a pen-type holder using either a micro-screw or by soldering.

4.4.1.3 Shaping a double-line chisel point
The sequence of operations is illustrated by numbers ① − ⑥: ① after tempering a drill rod of a specific diameter, for instance 1.2 mm, it is

clamped in a vice; ②, ③ and ④ both sides and also the centre of the point are ground using a micro-file; ⑤ and ⑥ the point is further ground and the sides and top are shaped; it is then re-tempered as was described in 4.4.1.2.

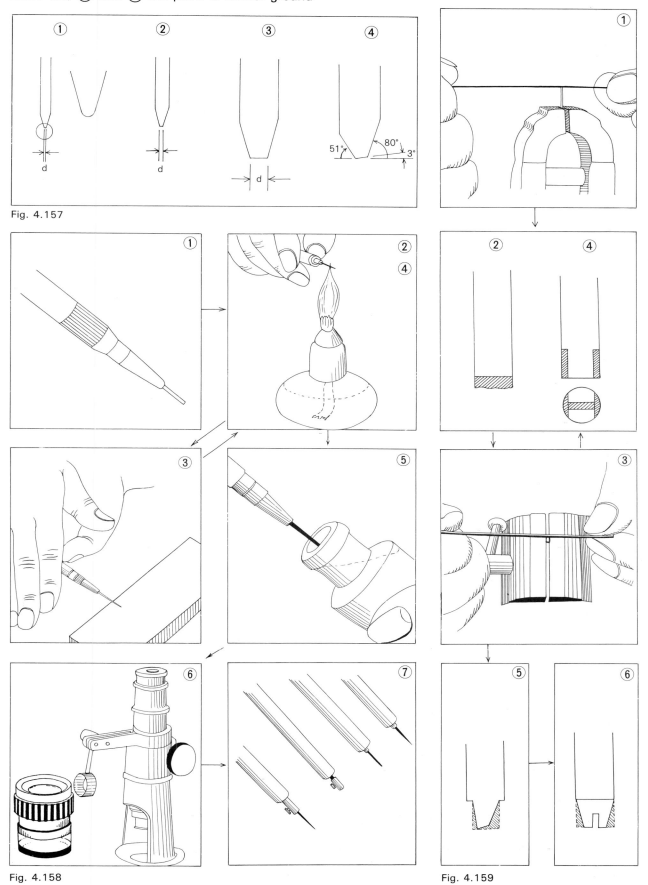

Fig. 4.157

Fig. 4.158

Fig. 4.159

143

4.4.2 The pen-type scriber

4.4.2.1 Scribing method
Figure 4.160. ① The point is held vertical to the scribing surface and uniform linework is drawn along the edge of a short, plastic ruler; ② the instrument should never be applied at an oblique angle, and care must be taken not to damage the surface of the base material. The point should only break the coating which is then removed. If the base is damaged, this disturbs the passage of incident light and will result in loss of line sharpness after contact exposure; ③ carefully use a plastic sponge to dust off flakes of coating removed during scribing.

Figure 4.161. Spring-loaded, pen-type scriber. This is an example of a commercially produced instrument. It incorporates a spring, enabling the application of uniform pressure on the coating, and results in good scribed linework. ① Variable tension spring; ② magnifying glass; ③ scribing point; ④ supporting legs.

4.4.2.2 Scribing parallel straight lines
① Resultant pattern; ② producing linework with equipment similar to that described in paragraph 4.2.4.1.

4.4.2.3 Producing numerals
These are drawn using a specially designed point and a template. ① The sequence of operations involved in scribing the elements of a figure. The example relates to a '2'; ② a standard point should not be used in association with a template; ③ specially designed 'hook points' and templates must be used when scribing numerals; ④ the shape and dimensions of a point.

4.4.2.4 Dotted lines
① If a dot scriber is not available, a continuous, solid line is drawn and then broken by the application of opaquing fluid; ② an example of a purpose-built dot scriber. The top of the cylinder is depressed, and causes the point to rotate and create a dot.

4.4.2.5 Another type of dot scriber
This incorporates a small electric motor which rotates a chisel point to scribe a dot.

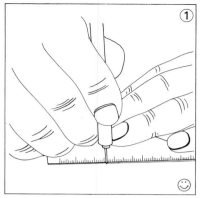
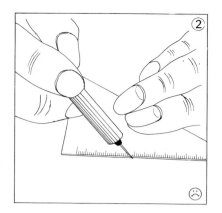
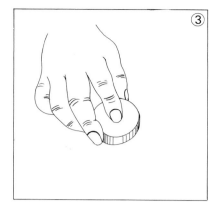

Fig. 4.160

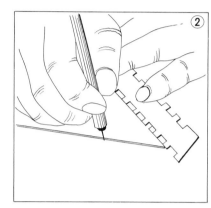

Fig. 4.161

Fig. 4.162

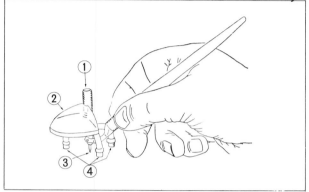
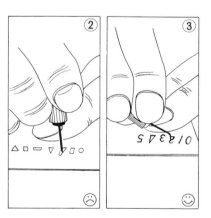
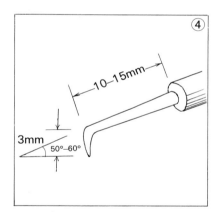

Fig. 4.163

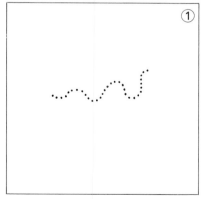

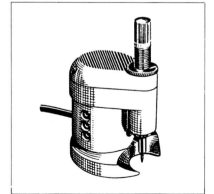

Fig. 4.164

Fig. 4.165

4.4.3 Tripod scribers

4.4.3.1 Producing regular width linework using a tripod scriber

Figure 4.166. Ready-made points. (a) A steel gramophone needle for scribing a single line; (b) a tungsten-carbide tipped steel needle for a single line; (c) ditto for a double line; (d) a sapphire-tipped steel needle for a single line; (e) ditto for a double line; (f) side view of a chisel point of the type used in the instrument shown in Fig. 4.171; (g) front view of the point described in (f); (h) ditto for a double line; (i) blade-type chisel points used in tripods originally designed in the USA.

Figure 4.167. The original fixed-head scriber tripod.

Figure 4.168. A spring-loaded, fixed-head scriber tripod. ① Variable tension spring; ② adjustment screw; ③ lever for raising or lowering the point; ④ magnifying glass; ⑤ scribing point; ⑥ supporting leg.

4.4.3.2 Double line scribing

Figure 4.169. Double lines produced using a swivel-head tripod. A pecked single or double line is initially scribed as continuous and then broken by applying opaquing fluid at appropriate intervals.

Figure 4.170. Swivel-head scriber tripod.

Figure 4.171. Cylindrical, swivel-head scriber tripod. ① Eccentric chuck; ② magnifying glass; 3 support stand used when inserting a chisel point.

Figure 4.172. Cylindrical, spring-loaded, swivel-head scriber tripod. ① Variable tension spring; ② magnifying glass; ③ scribing point; ④ adjustment screw; ⑤ supporting leg.

4.4.3.3 Using the original type of fixed-head tripod

① Insert a scribing point into the holder and tighten the clamping screw to prevent movement; ② mount the holder in the tripod and tighten the adjustment screw; ③ change the position of the point, in relation to a flat surface, by using the adjustment screw; ④ remove the front supporting leg; ⑤ hold the scriber between two fingers; ⑥ do not hold the top whilst scribing with a light, uniform pressure; care must be taken in order not to damage the surface of the base material.

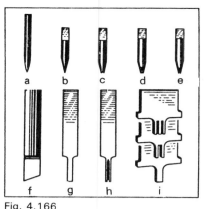

Fig. 4.166

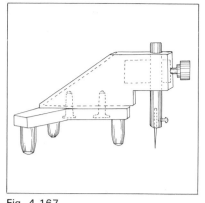

Fig. 4.167

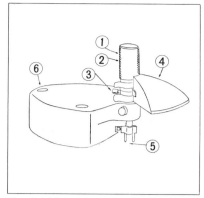

Fig. 4.168

Fig. 4.169

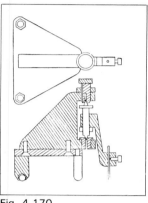

Fig. 4.170

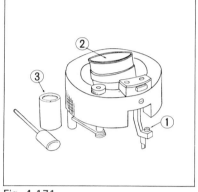

Fig. 4.171

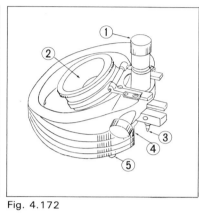

Fig. 4.172

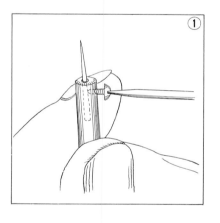

Fig. 4.173

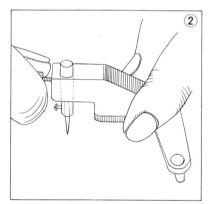

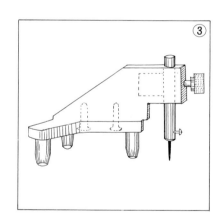

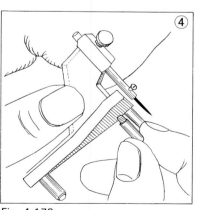

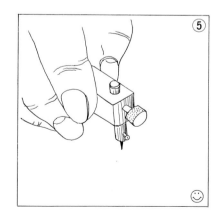

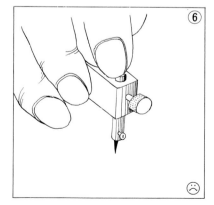

4.4.4 Correction and revision processes

Currently employed methods can be classified with respect to those which are 'direct', involve the 'photo-etching of pre-sensitised scribecoat', or require 'compound photography'. Which of these methods is employed, or considered as the most appropriate, depends on the amount or type of correction and/or revision necessary.

4.4.4.1 *The direct method for minor corrections or changes*

In the illustration: ① The original scribecoat which sometimes has a new guide image printed down onto it; ② incorrect detail is deleted using opaquing fluid. This may take place after essential new linework has been scribed in order to avoid disturbing the masking of an area which has been painted out; ③ corrections are scribed following either a new guide print down or a corrected drawing positioned beneath the original sheet.

4.4.4.2 *Linework requiring deletion*

This is painted out on the original scribecoat using a fine brush.

4.4.4.3 *Photo-etching pre-sensitised scribecoat*

This method is used to carry out the revision of detail in complicated areas or where linework is dense. It should be noted that line quality tends to diminish after the process has been repeated three or four times.

① Contact photography is used to print the original image onto scribecoat. K & E's Etchscribe pre-sensitised film is an example of a commercially available product; ② this is developed to produce a positive original, or reverse developed to generate a negative; ③ etching; ④ the material is dried, leaving a negative version of the original linework on the scribecoat; ⑤ new detail is then added.

4.4.4.4 *The compound photography technique*

This process is particularly suitable for the amendment of detail in large, built-up areas.

① Original detail; ② draft of revised content; ③ – ⑥ opaquing liquid is used to delete areas on the original, colour separated, scribe sheets that are at variance with the detail included in ②; ⑦ new scribecoat with pre-punched registration holes; ⑧ guide copy images of detail unchanged on original sheets ③ – ⑥, and from the revisions included on ②, for use in producing colour separations at ⑦; ⑨ lettering positive incorporating the stick-up of new information and omitting unnecessary or incorrect detail; ⑩ scribing of the new linework contained in ⑧; ⑪ a composite positive produced by the contacting of material shown in ③ – ⑥, together with ⑩. This is used with ⑨ which serves as an overlay for checking purposes; ⑫ and ⑬ further checks—in this case preparation of a new, open-window negative mask is required; ⑭ the final positive.

4.4.4.5 *Direct revision*

This requires the re-scribing of detail on original drawings, and is a skilled process. It reduces the number of steps necessary to complete revision, and also the amount of material used. ⑦ and ⑩ are produced by the method described in 4.4.4.1, rather than by using steps ⑦, ⑧ and ⑩ as described in 4.4.4.4.

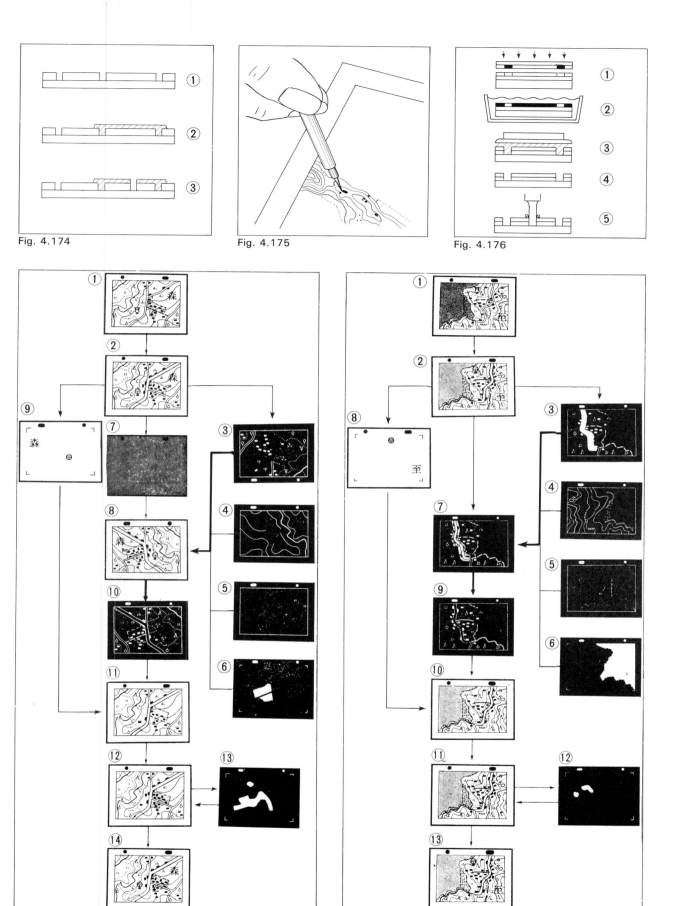

Fig. 4.174

Fig. 4.175

Fig. 4.176

Fig. 4.177

Fig. 4.178

4.4.5 Guide copy and multi-colour proof preparation

4.4.5.1 Positive to positive copying to produce a guide or key image

Figure 4.179. The principle of positive to positive copying, or the positive working process, involves the transfer of a right-reading image of a map compilation, drawn on a transparent base material, onto a sheet of scribecoat. This can be done by copying through the base to produce a right-reading version of it on the surface of the sensitised scribecoat. Face to face copying is undertaken when a wrong-reading scribing is required.

Figure 4.180. Use of the dry diazo method, and associated ammonia vapour development in a fume cupboard, is standard practice in positive working. The sensitive emulsion consists of a diazo compound and a dye coupler, and the process does not require darkroom use because of the low light-sensitivity of the diazo compound. The stages of the operation are illustrated from ① to ③: ① the diazo compound and dye coupler are mixed, and then wiped evenly over the clean surface of the base material using a soft pad. This is dried with a fan and then, together with an original document, placed in a vacuum frame and contact copied using an ultraviolet (UV) light source; ② a fume cupboard incorporating a fan and ducted ventilation system is essential during subsequent development in ammonia vapour; ③ completion of the simple process.

4.4.5.2 Negative to positive copying to produce a guide or key image

Figure 4.181. This technique, the negative working process, can be used to produce a guide copy on a sheet of scribecoat. It is a superimposure method, and the most common version involves water-soluble (but non-light-sensitive) photopolymers being mixed with pigments. Resultant colour depends on the pigment used, but there are a number of commercially available pre-prepared emulsions.

Figure 4.182. The superimposure method is illustrated from ① to ⑤. A darkroom is not necessary because of the low light-sensitivity of the photopolymer. ① Coat the base sheet with emulsion; ② dry the surface using a fan; ③ contact expose; ④ develop by washing under running water; ⑤ dry but do not wipe. Additional colours can be applied, as necessary, by repeating the process.

4.4.5.3 Using the superimposure process in the drawing office

① Prepare an appropriate amount of emulsion and mix this with a specified colour pigment. Evenly coat the surface of a clean plastic sheet by wiping the mixture over it with a soft pad;

② uneven and/or thick coating will result in the generation of a poor image at a later stage. In this case the coating should be washed off under running water and step ① repeated. Successful coating requires considerable skill, and it may be preferable to apply the mixture by using a whirler; ③ dry the coated surface with a fan, and then contact expose a negative original with it using a UV light source and a vacuum frame. A secluded corner of the room should be used when doing this; ④ develop with running water and by gentle wiping with a sponge; ⑤ poor coating and/or insufficient exposure time can cause the image to wash off during development. Thick coating and/or over-exposure may make it difficult to wash off coating from non-image areas; ⑥ gently blot the surface to remove excess moisture and dry with a fan.

4.4.5.4 Preparation of a multi-colour proof

Photomechanical proofing, also called pre-press or off-press proofing, is widely used to check the appearance of sets of colour separations in the drawing office. It is employed instead of press-proofing with ink which is both expensive and time consuming. Proofing in colour, i.e. checking the accuracy of registration between the different map components, is of greater importance than colour fidelity. It is accomplished by using individual colour overlay proofs, prepared from pre-sensitised transparent film, or by the superimposure method. The latter, which was described in 4.4.5.2, is popular because it is simpler and cheaper than other methods. Another technique, a transfer method marketed under the tradename Cromalin, is dry working and uses colour-toning powders. It is useful for the proving of colours and as a guide to their fidelity, but it is considerably more expensive than the superimposure process described above.

4.5 Masks, stick-up and dry-transfer

4.5.1 Masks

Symbols used to denote areas on maps are comprised of solids, tints and patterns, or combinations of these. They are used to indicate feature types and classes, but can also demonstrate differences in value or quantity. Areal variations represented by modifications of tone, for example shading and vignetting, are illustrated in subsections 4.3.5 to 4.3.8.

Masks for areas containing either solid colours, screened tints or patterns can be prepared by

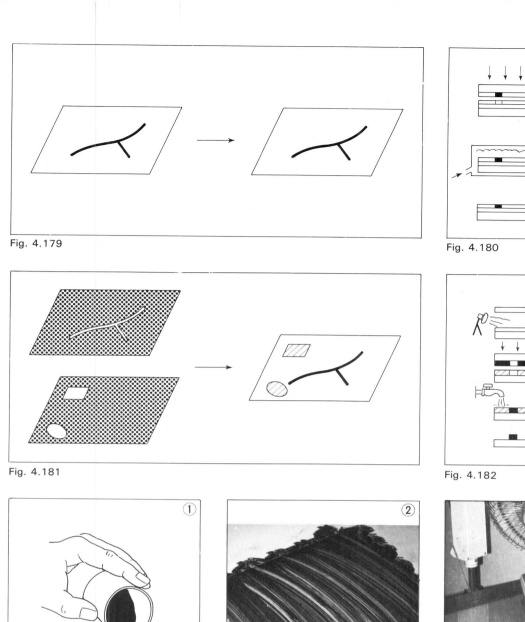

Fig. 4.179

Fig. 4.180

Fig. 4.181

Fig. 4.182

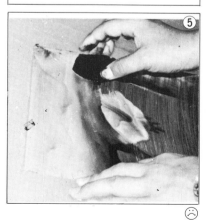

Fig. 4.183

151

several methods. Subsequently, specified symbol patterns, which are produced on strip-film, are mounted in relevant areas or a sheet-sized pattern or screen is used in combination with a mask when it is exposed during the manufacture of pre-sensitised printing plates.

Two kinds of masks, negative or positive, can be prepared to cover the full extent and shape of areas on map sheets. The negative version demonstrates 'open-window' image areas, and the positive type 'open-window', non-image detail.

4.5.1.1 Preparing small or simple masks

① A positive mask of a small solid area can be drawn as part of a line image, simply by filling in its outline with ink. That used for linework has been deliberately manufactured to be suitable for the generation of fine detail, and to flow easily from a pen. However, these inks often prove unsatisfactory when used in an attempt to fill in large areas. It is preferable to paint these with an opaquing liquid, of the sort used by photographers to correct negatives, in order to produce masks of areas of significant extent. ② Masking tapes are produced commercially and typically consist of self-adhesive, opaque, coloured plastic or paper. They are used to produce negative masks of large but regular shapes, such as the edges of a negative sheet, and prevent unwanted detail exposure in these places. ③ Commercially produced opaque paper. Pieces of this are employed to create negative masks during the production of printing plates. The material can also be positioned to create appropriate open windows during exposure.

4.5.1.2 'Cut and peel' method

The employed material consists of a polyester film base overlaid by a transparent, pliable, actinically opaque red membrane which is easily removable. It can be superimposed upon a map document, and enables linework comprising the latter to be followed with a cutter. ① The strippable surface layer should be cut using a scalpel or swivel-head scriber holding a needle or sharp blade. Cutting is not possible using a fixed-head scriber. ② Subsequently, unrequired areas of the red membrane can be peeled from the supporting base using either tweezers or a piece of adhesive tape.

As with all manual masking processes, great care must be taken to follow a guide image during the cutting process as, normally, the eventual tint will be required to fit to a bounding line or butt-join with another. Consequently, the manual cutting of a mask is only really an effective method when dealing with relatively simply shaped areas.

4.5.1.3 Pre-sensitised peel-coat

This material is commercially manufactured, and is normally used to produce masks relating to

details appearing in complex areas where precise registration is essential. Examples are provided by coasts and double-line river courses, etc. It differs from those used in operating manual methods in that the areas to be masked are directly derived, by photomechanical means, from the outlines scribed as part of the original drawings.

① A scribed or original photographic negative is exposed in a vacuum frame with the emulsion side of the negative in direct contact with the pre-sensitised layer on the top of the peel-coat; ② development: the side exposed is placed face down in developer contained in a chemically-inert tray which is slightly larger, in all directions, than the sheet of material; ③ washing: after development has been completed, the solution is rinsed from the surface of the peel-coat with tap water; ④ drying: remaining surface water can be removed with a squeegee or blotting paper; ⑤ etching: after about another hour an applicator block is soaked with etch solution and gently wiped over the developed, pre-sensitised surface using light pressure. It should be allowed to work on its own and not rubbed in. The operation is repeated, and the exposed image area and its membrane are removed to leave the unexposed background in the form of a stencil; ⑥ and ⑦ the etch solution is then flushed off with water, and the material is squeegeed or blotted as in ④; ⑧ peeling takes place as was described by ② in paragraph 4.5.1.2, and unrequired lines or areas are deleted using an opaquing liquid or dye solution which is applied with a brush.

If a positive-working pre-sensitised material is used, the exposure and developing sequence is virtually the same, except that the positive original (in the case of some materials) is exposed through the base of the peel-coat. This leads to a slight reduction of linework width on the negative mask, and contributes to improved colour registration during printing.

4.5.1.4 Etched peel-coat images

When using this technique it is necessary to opaque out and fill in all unrequired linework before peeling. This results in the production of slightly smaller peeled areas due to the deletion of the width of bounding lines from the eventual masks, as can be seen in ①. If this was not the case, the areas would be bigger, as in ②.

4.5.1.5 Successive peeling of areas

For reasons of economy, a single peel-coat can be employed to produce a number of masks. For example, when preparing hypsometric or bathymetric tints the successive peeling of each individual layer, and its immediate contacting to produce a positive, saves material. Eventually the series of positives so generated is combined to provide a single colour component.

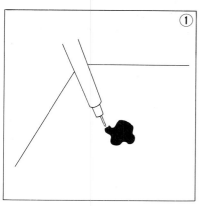 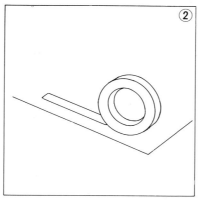 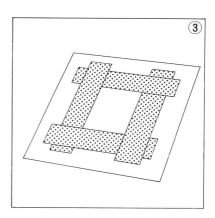

Fig. 4.184 Preparing small or simple masks

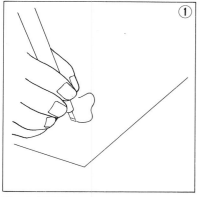 

Fig. 4.185 'Cut and peel' method

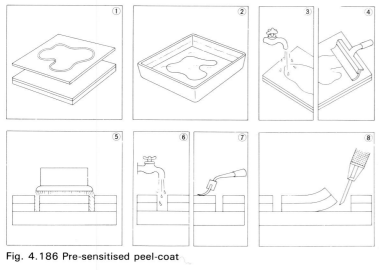

Fig. 4.186 Pre-sensitised peel-coat

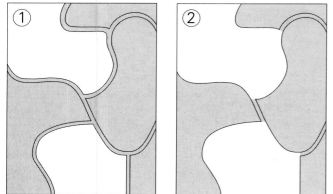

Fig. 4.187 Etched peel-coat images

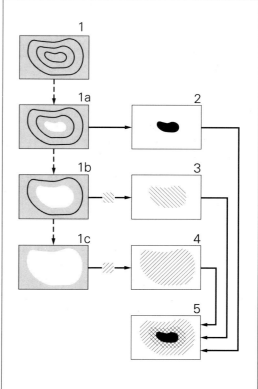

Fig. 4.188 Successive peeling of areas

4.5.2 Stick-up

Currently this is the normal method employed for the manual generation of a work-sheet containing lettering, point symbols, etc. The preparation process involves the mounting of names and other devices, produced on film or paper, in position on the work-sheet relating to a specific colour with an adhesive medium. Stripping film is normally used and consists of a thin membrane, with a sensitised emulsion, which is carried by a standard thickness film base. The complete sheet, incorporating the required lettering and symbols, results from its exposure using a phototypesetter and subsequent development to produce a right-reading positive. An adhesive solution appropriate for the small pieces of stick-up can be prepared by mixing acetone (91%), lactic ethyl (6%) and glacial acetic acid (3%).

4.5.2.1 The stick-up process
① Cut the membrane around each name or symbol using a scalpel or knife of the type described in 4.2.5.13; ② and ③ peel each of the symbols from the backing film using a cutting pen and tweezers; ④ the membrane is both thin and delicate, and care must be taken not to damage it when picking it up; ⑤ after moistening the underside by painting on adhesive with a small brush, the image is appropriately positioned with the tweezers; ⑥ blotting paper is put over it and absorbs the excess solution which is forced out by exerting light pressure with a blunt instrument.

4.5.2.2 The results of stick-up
① good; ② uneven positioning; ③ the use of too much adhesive damages the transparency of the work-sheet.

4.5.2.3 Wax coating
This can be applied rather than an adhesive solution. It is normally used for larger sizes of lettering which may be employed for titles, etc. Some pre-waxed stripping films are commercially available.

Figure 4.191. Peel the membrane from the backing film using tweezers and mount it, face down, on a plastic sheet.

Figure 4.192. Coat the rear of the membrane (the wrong-reading side) with a thin layer of wax, either by painting this on manually or using an electric waxer.

Figure 4.193. Cover the wax with a protective backing sheet: ① the right-reading film membrane; ② the wax coating; ③ the backing sheet.

4.5.2.4 Cutting between letters enables type to be curved
① Cut slits at right angles to the direction of the type; ② – ④ position the first letter at the designated starting point and follow this with each of the others working along the desired curve.

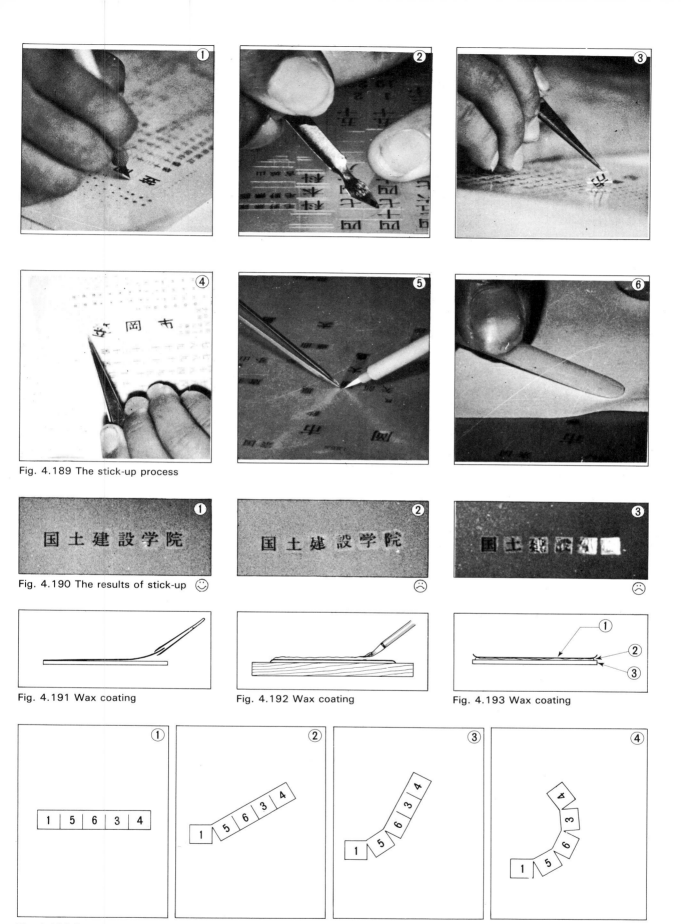

Fig. 4.189 The stick-up process

Fig. 4.190 The results of stick-up ☺ ☹ ☹

Fig. 4.191 Wax coating

Fig. 4.192 Wax coating

Fig. 4.193 Wax coating

Fig. 4.194 Cutting between letters to enable curving

155

4.5.3 Dry transfer

The use of pre-printed dry transfer materials makes possible the application of positive lettering and symbols to various map components. In addition, it enables the combination of linework with dot screens, and other patterns and rulings, without the need for complex masking and photomechanical operations.

4.5.3.1 Commercially available materials

These are produced in many type styles and as various symbols and patterns. The sheets exhibit a relatively small format and, in consequence, the available screens and tones are normally only used for small areas. Pre-sensitised films are also manufactured for the purpose of allowing the creation of personally produced dry transfers.

4.5.3.2 Working method

1 An individual letter or symbol is transferred onto the surface of the work-sheet by gently rubbing the front of its carrier sheet with a blunt instrument or soft pencil; 2 the sheet is then pulled away and the letter remains on the work-sheet. Mistakes can easily be corrected by gently pressing a piece of adhesive tape on the incorrect letter and lifting it from the surface. When completed, the work-sheet can be sprayed with a clear, protective coating which helps to protect it from damage during handling or storage.

4.6 Lettering in Roman characters

As is the case with all graphic icons appearing on the face of a map, letters are a method of symbolising and encoding details representing geographical reality. However, the most straightforward use of characters occurs when they are employed as literal symbols, in combination with variations in their styles, sizes and positions, to provide information about other graphically displayed elements. Thus they serve to demonstrate differences in the names of individual places, regions and other spatial features, whilst also giving numerical details, map titles and legend descriptions, etc.

The cartographic quality and appearance of a map is very much dependent on the way in which included lettering has been employed. Usefulness of a reference map is directly related to the clarity and legibility exhibited by the type styles used and the positioning of the lettering. Distinction and recognition of a feature to which a name label applies; the search-time necessary to find names; and the ease with which letters can be read are all important to the functions of a map and its success or failure as a medium for the communication of spatial information to a user.

Problems relating to the specification of type, and

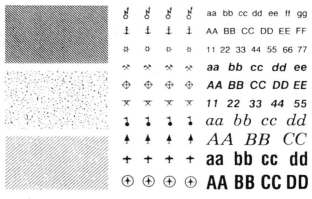

Fig. 4.195 Commercially available materials

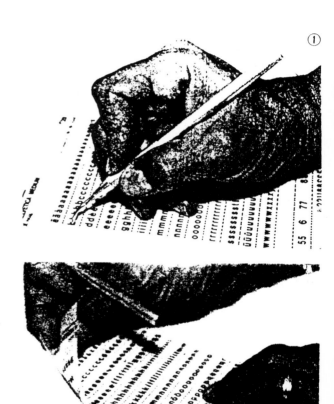

Fig. 4.196 Working method

the positioning of names, are discussed in *Basic Cartography* Volume 2 with respect to their role in map compilation. Methods of character generation are also mentioned in this chapter as being classified as of four kinds: stick-up and dry transfer; mechanically produced; freehand; and computer-assisted lettering. The first of these currently demonstrate significant advantages over other methods, in that they are generally quicker and require less drawing skill. However, their fundamental drawback is that there are, necessarily, restrictions in terms of both type styles and sizes as compared with the versatility inherent in good quality hand lettering. Stick-up and dry transfer are explained in subsection 4.5.2, and photo-typesetting and computer-assisted production are considered in Section 4.7 with respect to their use in creating Sino-Japanese characters.

Freehand lettering. This is the most flexible method available for the application of names to a map. It is highly desirable that all cartographers become reasonably adept at the freehand production of characters, in order that the significant amounts of lettering necessary during the compilation process will be neatly executed. Typically, the layout of titles, legends, etc., will be required, and also the computation of letter spacing, during the design process prior to the obtaining of stick-up materials. Individuals need to gain a facility in three vital aspects: spacing; the use of guidelines; and stroke technique. In the following paragraphs explanations are provided relating to Gothic and Roman characters, which are the two most commonly used sorts of cartographic type styles.

The sequence of strokes and use of guide-lines in producing Gothic letters are explained in paragraph 4.6.1; the width, spacing guide-lines, and generation of Roman characters are demonstrated in 4.6.2, and those for Gothic capitals in 4.6.3. Mechanical lettering procedures are detailed in 4.6.4.

4.6.1 Gothic, upper and lower case

Single-stroke lettering is described here, and consists of characters in which the width of all linework is equivalent to the thickness of the point of the pen with which it is drawn. The method is used almost exclusively for letters which are less than 8 mm in height. It is the easiest lettering style to produce, and also the quickest.

4.6.1.1 Basic techniques
Freehand strokes consist of those which are drawn straight down, at an inclined angle, horizontally from left to right, and curving from top to bottom. Up strokes cannot be made with a sharp pen or pencil unless an aid such as a glass ruler is employed.

4.6.1.2 Characters consisting of straight lines
Included arrows and numbers demonstrate the direction and order in which various parts of letters are drawn, and are related to the grid squares which provide guide-lines for character construction. Once the correct method of making composing strokes becomes a habit, better lettering will result, regardless of the tools available.

4.6.1.3 Characters consisting of curved lines
The remarks made in 4.6.1.2 are again relevant.

4.6.1.4 Lower case characters
Comments made in 4.6.1.2 pertain, but horizontal guide-lines are particularly important. These consist of: a 'cap' line (for ascenders), 'waist' line, 'base' line and 'drop' line (for descenders), and can be easily prepared in pencil with a plastic device such as the letter spacing template illustrated in Fig. 4.43.

4.6.1.5 Numerals
Remarks in 4.6.1.2 and 4.6.1.3 also apply to numerals.

4.6.1.6 Italic characters
These slanted or inclined Gothic letters are constructed at an angle of 70° to a horizontal base line. The space ratios between guide-lines are the same as those depicted in Fig. 4.200.

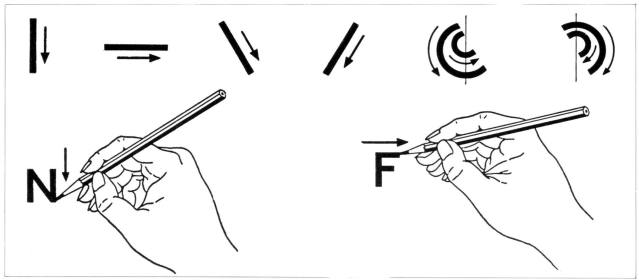

Fig. 4.197 Basic techniques

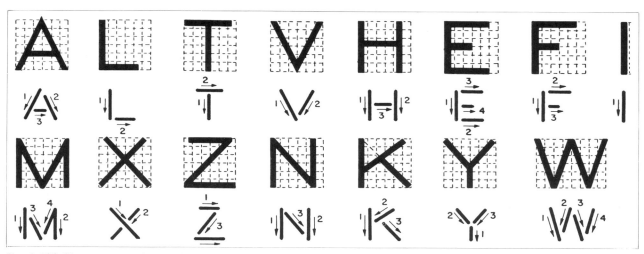

Fig. 4.198 Characters consisting of straight lines

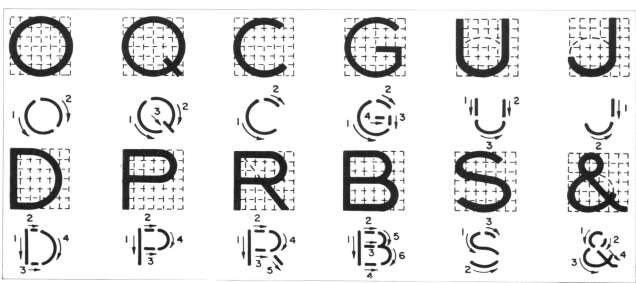

Fig. 4.199 Characters consisting of curved lines

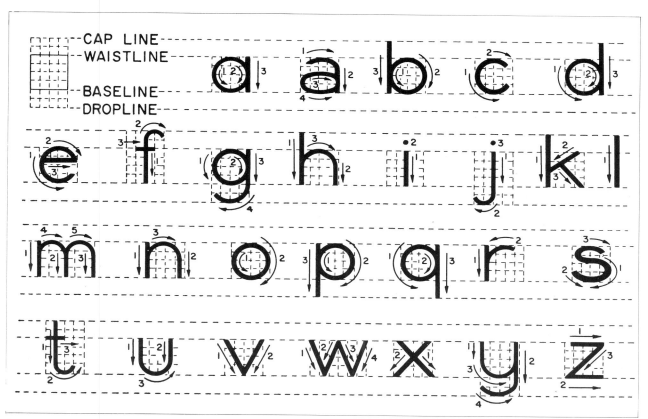

Fig. 4.200 Lower case characters

Fig. 4.201 Numerals

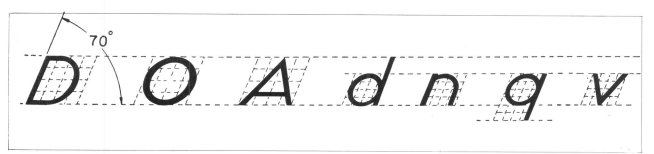

Fig. 4.202 Italic characters

159

4.6.2 Roman characters, spacing and letter construction

4.6.2.1 Letter spacing

In order to achieve the correct spacing between letters the net dimension of the width of each printed character is employed. This amount is determined graphically by finding the vertical centre line of the overall area of every letter of a specific type style's alphabet, and then allowing equal spacing between the net widths of each. In the words 'NEW YORK' the letters 'N', 'E' and 'W' exhibit overall/gross widths of 7, 6¼ and 10 units respectively. However, their net dimensions are 5, 4¼ and 6¾ units. A uniform distance of 3½ units is selected as the gap between the net values of the letters, which are then spaced accordingly. This method is not really practical if characters are less than 10 mm in height, but careful consideration of the principles involved will help to develop an artistic appreciation and visually correct spacing.

4.6.2.2 Components of Roman characters

The width of the thicker body strokes should be equivalent to between 1/6 and 1/8 of the height of a letter, and the thinner or 'hair lines' significantly narrower by comparison. The sequence and direction of strokes used in drawing letters are indicated, respectively, by numbers and arrows. Good and poor examples are provided of the small curves or 'fillets' occurring at the 'serifs' (ends of lines). In the illustration these have been produced by equal horizontal extension of a line on both sides of the upright. Fillets are then drawn to connect the ends of this new line to the main stroke.

4.6.2.3 Drawing Roman italic characters

Construct letters in the same way and with equivalent proportions to upright Roman characters, but incline their axes as in single-stroke italic lettering. ① Roman italic upper case; ② Roman italic lower case; ③ Roman italic numerals.

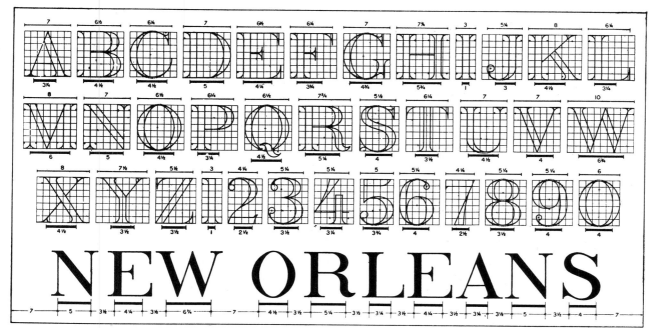

Fig. 4.203 Spacing

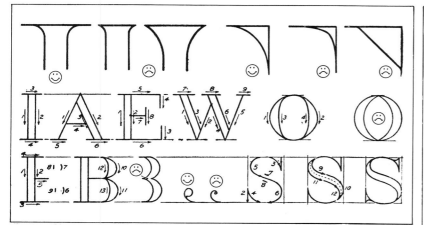

Fig. 4.204 Components of Roman characters

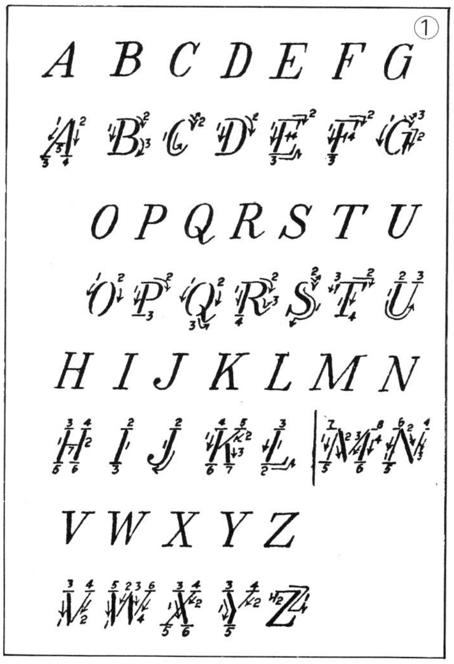

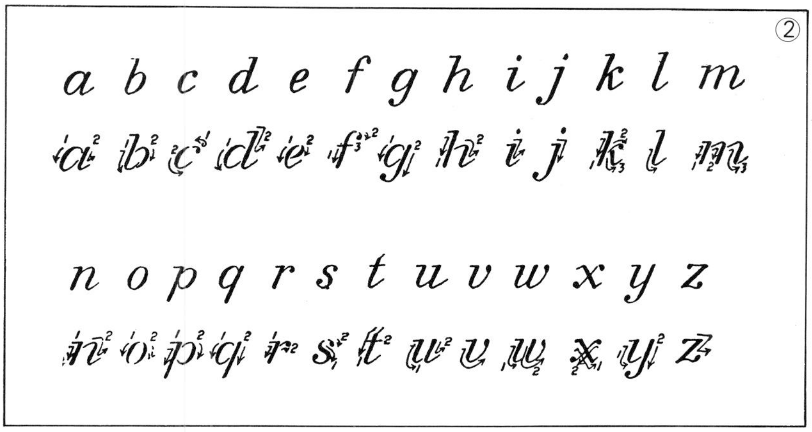

Fig. 4.205 Drawing Roman italic characters

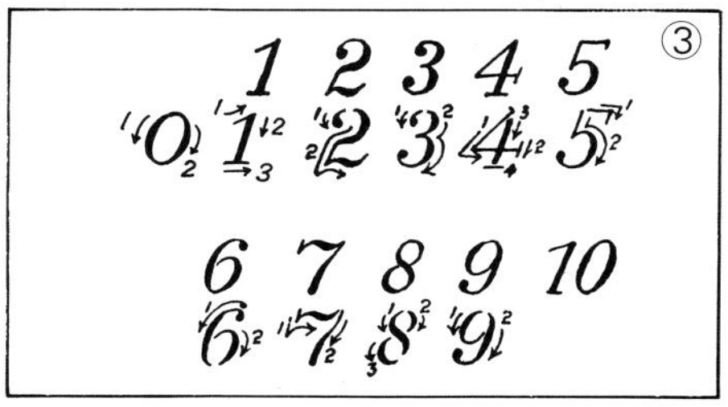

161

4.6.3 Spacing Gothic upper case

4.6.3.1 Spacing
The technique employed is the same as that described in 4.6.2.1 with reference to the Roman type style. However, as is shown in the illustration, the net width dimensions and spacing are different with respect to letters of the same height in the other alphabet.

4.6.3.2 Drawing Gothic characters
The outlines of letters are produced in the same way as upright, single-stroke letters, but by using two strokes rather than one. The width of lines comprising finished, standard lettering is approximately 1/7 of the height of a capital. Proportional width is indicated by a number related to a height of 6 units. The sequence of operations for drawing linework is shown. Wide letters such as 'M' should be finished with a slight spur.

4.6.4 Mechanical lettering

Neat and acceptable type can be produced by mechanical means. However, the characters produced by most devices give their reader an impression of viewing the lettering appearing on a construction engineer's blueprint. The equipment is particularly useful in the preparation of engineering drawings where variations in the style and orientation of lettering are not essential. It is likely that mechanical aids of this sort will continue to be important in 'Do It Yourself' cartography, especially for the production of graphs, charts and cartograms.

The most familiar equipment consists of a special pen which releases ink from a reservoir through a narrow tubular nib whilst being guided, either mechanically or manually, with the aid of a template.

1 **Wrico**: a patented name for a lettering system requiring the use of perforated stencils (or guides) and special pens. A relevant guide is positioned directly upon an area to be annotated, and is moved backwards and forwards in order to facilitate the following of letters necessary to form a word or label. The pen is held in the hand and moved around the letter shapes contained by the stencil which is kept slightly above the working surface by blocks which prevent smearing. A considerable variety of type styles is available, including both 'extended' and 'condensed' versions, but a separate stencil is necessary for different sizes, although pen thicknesses can be changed.

2 **Variograph**: is also a patented brand name relating to a very versatile system involving the employment of a template incorporating incised characters and a stylus. Essentially, the instrument used is a small, adjustable pantograph which fits into the lettering guide. Letters are traced from the template and reproduced in ink by a pen located at the opposite end of the pantograph to the stylus. Adjustments can be carried out in order to modify letter size and style (i.e. extended or condensed), using merely one template. The latter are available in various type faces.

3 **Leroy**, another patented system requiring use of templates, a scriber and dedicated drawing points. Different templates are necessary to produce size and style variations in resultant lettering. An appropriate template is moved along a T-square or straight-edge whilst the scriber is employed to trace letters incised into it. Simultaneously, individual letters are reproduced by the pen which works at a distance from the template. It is possible to generate many lettering styles and sizes, in both capitals and lower case, by interchanging templates and/or pens.

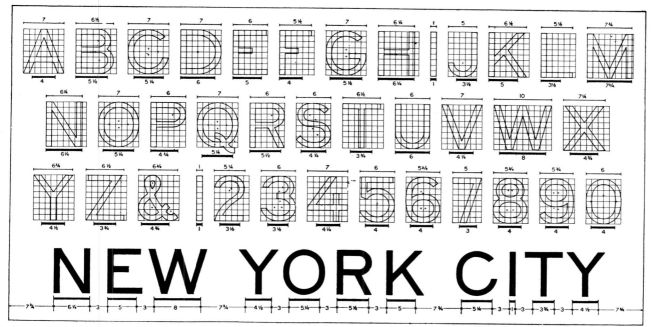

Fig. 4.206 Spacing

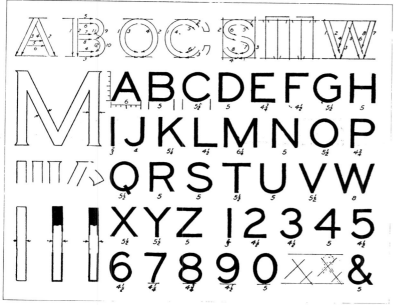

Fig. 4.207 Drawing Gothic characters

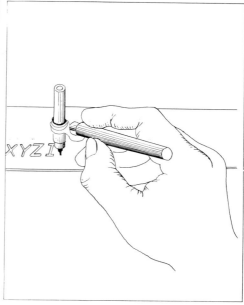

Fig. 4.208(a)

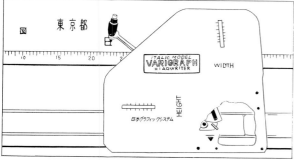

Fig. 4.208(b)

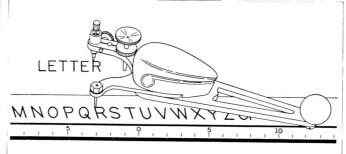

Fig. 4.208(c)

4.7 Lettering in Sino-Japanese characters

Sino-Japanese characters are discussed in this section as an example of the use of a national language on maps published for domestic purposes rather than international use. Written Chinese consists solely of ideographic characters, but Japanese also employs phonetic symbols. Chinese Kanji ideographs are used, but pronounced differently, in conjunction with Kana, which is a Japanese syllabary comprised of fifty phonetic sounds which are written in two types of characters. The first of these is Hiragana, which is a cursive syllabary, conventionally used for assisted words such as prepositions, adverbs, etc. The other, Katakana, is the square form syllabary used in the transliteration of unfamiliar words or foreign place names on a map. Kana was originally derived from Kanji and employs the same pronunciation. Hiragana was developed from the Kanji cursive script, and it too uses curved lines. Additionally, Katakana also originated from part of the Kanji square style and its characters consist primarily of straight lines.

4.7.1 Toosen style, proportions of principal characters

All strokes comprising Sino-Japanese characters have been designed to fit within squares which, when in combination, are regularly spaced.

4.7.1.1 Composition of strokes
As with freehand lettering, it is essential to ensure the correct composition of the strokes representing the various characters. Many of the horizontal or vertical parallel straight lines are equally spaced, and strokes are often positioned to be symmetrical on both sides of the square's centre line.

4.7.1.2 Composition of the elements of various characters
Good and poor stroke shapes have been used to represent essential elements of a number of characters. Good linework depicting fundamental aspects is shown by thick strokes within squares in the first and third of the rows. Poor examples of attempts to draw the same strokes are included as thin linework in the other rows.

Fig. 4.209 Composition of strokes

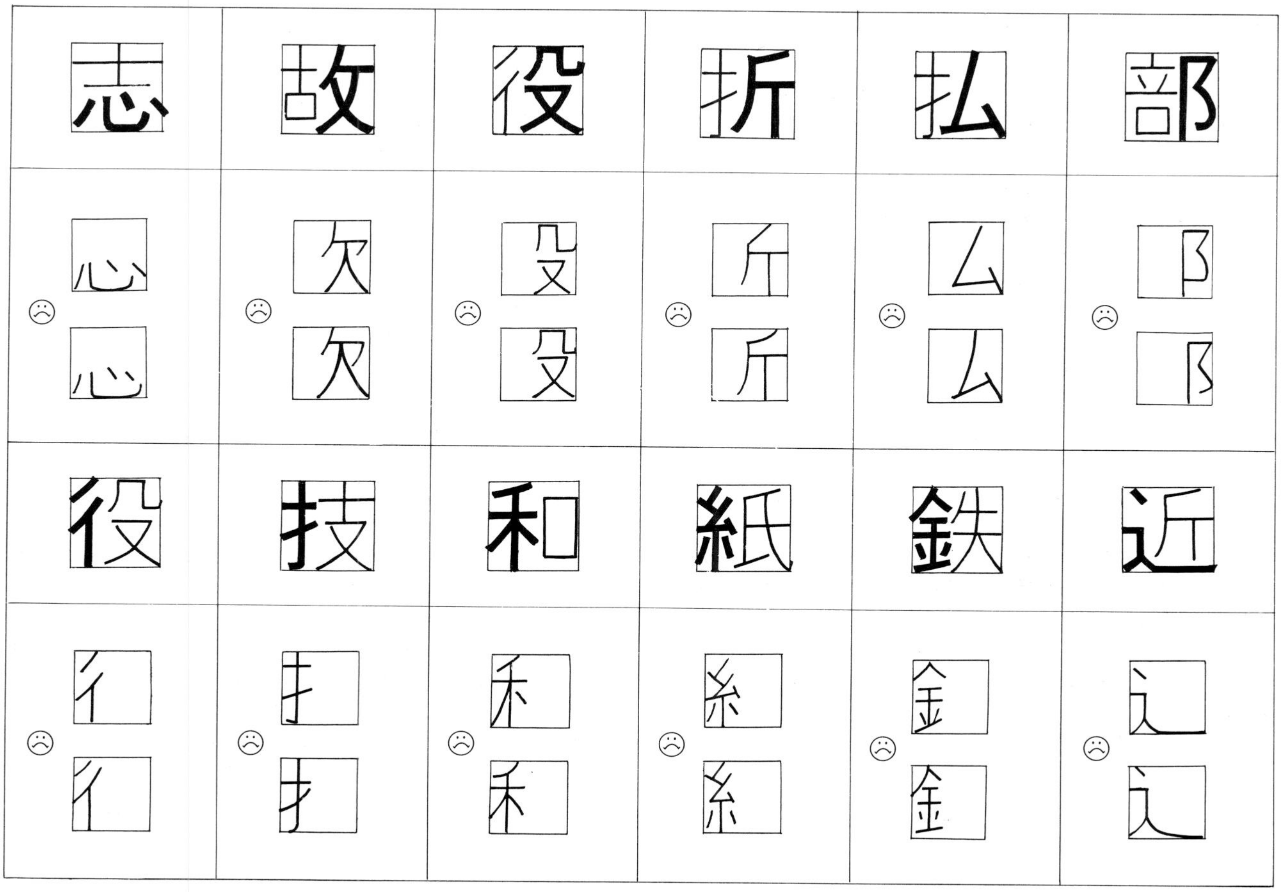

Fig. 4.210 Composition of the elements of various characters

Table 4.3

Style	Line width (1)	(2)	Legibility	Clarity	Remarks
Min-cho	0.2	0.9	A	C	Similar to Roman
Toosen	0.8	0.8	B	A	Similar to Gothic
Fantel	0.9	0.2	C	D	Horizontal lines thick, verticals thin
Rei-tai	—	—	D	B	Old style

Note: Quoted line widths indicate the ratios of horizontal (1) and vertical (2) thicknesses to letter height (10.0) in the standard/medium style. Versions inclined from left to right, together with other alternatives, are also available.

4.7.2 Styles of characters used on maps

4.7.2.1 Principal styles employed

The styles of standard characters regularly used on topographic maps are illustrated. ① Min-cho style with standard/medium thickness lines; ② Min-cho style with thick/bold lines; ③ Toosen style with thin/light lines; ④ Toosen style with standard/medium lines; ⑤ Toosen style with thick/bold lines; ⑥ Maru-Gothic style with standard/medium lines; ⑦ Maru-Gothic style with thick/bold lines.

The size of Kana. Roman characters and numerals is reduced to 80% of that of Kanji in order to obtain the same visual impression of height.

4.7.2.2 Styles occasionally used on maps

The peculiarities of the Rei-tai, Fantel and Maru-Gothic styles are as follows.

Rei-tai style: This was used on old maps, but is seldom employed today. The main methods for creating strokes are as shown: ① the beginning of a horizontal stroke should appear sharp; ② the tip of a vertical stroke should appear sharp; ③ horizontal strokes should have regular edges; ④ the base of the vertical stroke on the left should appear as sharp as that parallel to it on the right; ⑤ the central horizontal line should be directly connected to the right-hand vertical stroke; ⑥ the horizontal stroke should not terminate in a curl in case an inclined stroke with a curl also exists; ⑦ the end of the vertical stroke on the left should appear sharp, as does that on the other line parallel to it.

Fantel style: The distinguishing features of this recently introduced style are that vertical lines are all thin and horizontal ones thick. The appearance of the ends of strokes is specified as follows: ① the tip of a vertical stroke includes a tick, and inclined curves gradually increase in thickness towards their bases which appear sharp; ② an upward sloping stroke commences at the sharp edge of a thick line which gradually tapers along its course; ③ Rekka, a fundamental leg part of Kanji, is drawn as indicated; ④ Kagi and Otunyou are fundamental elements of Kanji. The correct shape of stroke ends is demonstrated by means of good and poor examples.

Maru-Gothic style: ① inner corners should be sharp.

① 東京都目黒区東山ちずチズＡＢＣ１２３
② 東京都目黒区東山ちずチズＡＢＣ１２３
③ 東京都目黒区東山ちずチズＡＢＣ１２３
④ 東京都目黒区東山ちずチズＡＢＣＩ２３
⑤ 東京都目黒区東山ちずチズＡＢＣ１２３
⑥ 東京都目黒区東山ちずチズＡＢＣ１２３
⑦ 東京都目黒区東山ちずチズＡＢＣＩ２３

Fig. 4.211 Principal styles employed

Fig. 4.212   Fig. 4.213   Fig. 4.214

167

4.7.3 Eiji-Happoo in Min-cho style

The principles relative to the creation of strokes used in the Min-cho character style are best exemplified with reference to Eiji-Happoo. Each of the steps involved in stroke generation is illustrated, at an enlarged scale, by both good and poor examples. Standard line widths, etc., for the main styles are tabulated below.

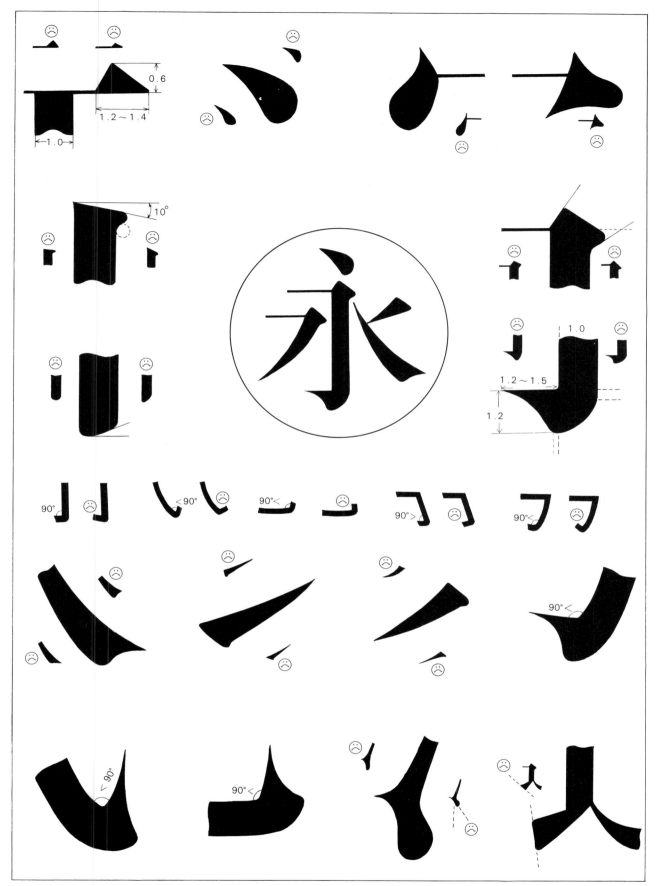

Fig. 4.215 Basic shapes of elements comprising parts of the Min-cho style

169

4.7.4 Alternative versions of names and their display (bilingual lettering)

4.7.4.1 General remarks and examples of the use of alternative languages

When using lettering relating to two languages, the character height employed for the second is normally reduced to between 60 and 80% of that employed for the main entry. If these details are displayed on separate lines, the original language appears on the first and the alternative below it. It is desirable to leave a space between them equivalent to at least 1.5 times the height of the type used to represent the second entry in the case of a name for a large area. In other instances the spacing may be reduced to the same as that of the lettering or less.

If providing annotation for a point symbol with two versions of a name, each consisting of a single line of letters, characters representing the main language should be positioned as close as possible to the relevant spot, and the alternative version slightly further away.

4.7.4.2 Transliterated names

When displaying transliterated versions of geographical names first recorded in another language, incorporation of the original version may sometimes be necessary. These secondary annotations should consist of smaller letters and be centred beneath the transliterated form of a name.

4.7.4.3 Multi-lingual names

It may be appropriate to incorporate multi-lingual lettering on maps of areas with more than one official language, or on those intended for international use. Normally this policy is particularly applied to detail included in legends, and geographical names appearing on the face of the map are in only one language form.

北 太 平 洋　神 奈 川 県

NORTH PACIFIC OCEAN　KANAGAWA

(87) ⊙ 煙突 CHIMNEY

△ 太郎山 Tarō Yama 123

△ 太郎山 Tarō Yᵃ 123

Tarō Yᵃ 太郎山 123 △

弁天島 Benten Sima

Benten Iwa 弁天岩

春川 *Haru Kawa*

Haru Kᵃ 春川

春 川 *Haru Kawa*

恵比須町 Ebisu

宝 町 Takara

竜飛埼 Tappi Sⁱ

竜飛埼 Tappi Sⁱ

小島 (25) Ko Sima

(25) 小島 Ko Sᵃ

Fig. 4.216 General remarks and examples

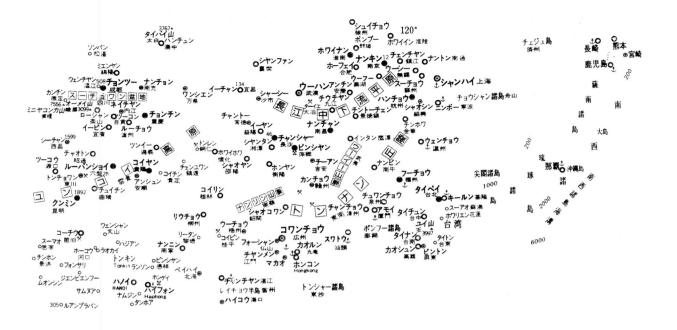

Fig. 4.217 Transliterated names

Höhe in Meter
Altitude in metres
Altitude en mètres

Emi Koussi
3415

Altitud en metros
Quota di altitudine in metri
الإرتفاع بالمتر

Fig. 4.218 Multi-lingual names

4.7.5 Phototypesetting and computer-assisted lettering

4.7.5.1 Phototypesetting

This process culminates in the generation of positive lettering or symbols on either photographic film or paper following the contact exposure of relevant negatives. After development, the results are manually mounted in their appropriate positions and aligned on a names work-sheet as was described in subsection 4.5.2. Typesetting is normally carried out using a list of names abstracted from the compilation document, all of which have been specified with reference to required sizes, styles, spacing and whether they are to appear in an italic, extended or condensed form, etc.

First generation machines: The setting of each character was accomplished by manual rotation of a font disk, or the sliding of a frame, and the exposure of letters or symbols held on photographic matrices. Original models were standard typewriters. A schematic layout is illustrated in the figure: ① light source; ② main frame of negative font matrix; 3 sizing lens; ④ modifying lens; ⑤ shutter; ⑥ sensitive film; ⑦ magazine; ⑧ letter monitoring window; ⑨ layout inspection indicator. Font frames were interchangeable, and

it was also possible to develop alternative ones. This made the equipment useful for the setting of type in languages such as Sino-Japanese which have many character forms, and also for the combination of multi-lingual lettering such as Sino-Japanese and Roman. First generation machines are still useful for producing small amounts of typesetting or supplementing the production of high-speed equipment.

Second generation machines: A schematic layout is illustrated in the figure: ① light source; ② font disk; ③ sizing lenses; ④ sensitive film. This equipment is capable of achieving a moderate output speed resulting from operation of a keyboard. Depression of a key immediately initiates the exposure of photographic film through a negative image of the required character. These icons are mounted on a rotating font disk that is stopped for each exposure. An in-built control system makes allowances for inter-letter spacing between exposures, and character size can be modified by a zoom or sizing lens. This procedure is termed 'on-line', and there is an alternative one which is 'off-line'. The latter involves accessing a disk on a separate keyboard which is electrically operated by a computer for further processing. This method is effective for the construction of a database of

171

names which can be called up for use with maps which are subject to frequent or regular revision.

Third generation machines: This type of equipment is noted for its ultra-high output speed which is necessary for the production of the large amounts of type required as continuous text or 'body matter' on the pages of a book. Its schematic layout is illustrated in the figure: ① memory; ② cathode ray tube (CRT); ③ lens; ④ sensitive film. Characters are stored in digital form on a disk, and are encoded as to type style and size. Each letter is generated by the programmed arrangement of minute squares, and is flashed onto a high-resolution CRT by operating a keyboard. This can be carried out either 'on-line' or 'off-line', and requires that characters are exposed, through a lens, onto photographic film which is passed in front of the luminous CRT screen.

Fourth generation machines: Equipment such as an image-setter is noted for its laser beam printer (LBP) with a high resolution of 1 000 dpi (dots per inch) or greater. The machine produces final film or camera-ready copy suitable for use in plate-making, and a single sheet can incorporate both lettering and graphics consisting of linework and symbols. Preparation of these materials can take place directly from original compilations (i.e. 'computer to plate') once the necessary database has been generated.

4.7.5.2 Computer-assisted lettering

The theory of the process enabling the computer production of material in camera-ready copy form is illustrated in the figure: ① flatbed scanner, with a resolution of at least 1 000 dpi, for the capture of data relating to the positions and alignments of lettering and graphics; ② personal computer for the entry of type attribute details such as style, size, special feature, etc.; ③ small LBP for the output of proof sheets; ④ image setter allowing production of final films or original drawings.

In this small system the lettering database is assembled by using a personal computer and includes attributes such as style, size, spacing, etc. The positioning and alignment of lettering on a drawing are carried out manually by the stick-up of stripping film produced by a computer-driven phototypesetter. Examples of these are shown in Figs 4.220 and 4.221. Final films, and ultimately printing plates, are made (including the icons mounted on the lettering work-sheet) by the normal methods. Alternatively, a name database is prepared by extracting relevant detail from a compilation using the method cited in 2 above.

The complete operation can be performed by using an appropriate 'Desk Top Publishing' (DTP) system. This is a computer-assisted, page-layout workstation with a 'WYSIWYG' function, a 'PDL', and a connection to an LBP. The acronym 'WYSIWYG' stands for 'What You See Is What You Get', thus any image displayed on a CRT can be output as hard copy; and 'PDL' is 'Page Description Language' which incorporates functions appropriate for specification of the description and layout of both relevant type and graphics. A typical PDL is 'PostScript' which has been developed by Adobe and can specify the styles and sizes of letters; their positions; enlargement or reduction; angle of inclination with respect to the horizontal; and the vertical, inclined and curved alignment of both characters and graphic materials. DTP can be employed for compilation, and the resultant output can constitute the final documents ultimately used in the making of printing plates. The cartographic application of DTP is normally referred to as 'Desk Top Mapping' (DTM).

4.7.5.3 Fonts

A font consists of icons of the same style and size. In order to cater for the enormous number of characters comprising Kanji, a special font, designated as Jis Kanji, has been developed (JIS is the Japanese Industrial Standard). It includes a total of 12 999 characters which are sufficient for the labelling of virtually all geographical features.

Phototypesetting fonts: A wide variety of fonts are currently available for use in phototypesetting. However, almost all of these were originally designed for use in the printing of continuous text/books. Because of this there is a tendency for lettering which appears to be of sufficient strength when used as text material to create a lighter impression when included on maps. Thus it may be necessary to strengthen it to achieve the same visual effect and maintain clarity and legibility. A comparison of the respective designs of composing elements, and the width ratios between vertical and horizontal strokes comprising Minchoo characters of the same size, but intended for different purposes, is provided in the figure: ① newspapers; ② normal phototypesetting; ③ topographic maps.

Digital fonts: These are comprised of characters resulting from computer processing, and can be classified into two sorts, as is demonstrated in the figure: ① bitmap or dot font composed of 16 × 16 dots; ② outline or vector font. The former consists of a set of pixels, whilst the latter is stored in the computer memory as data relating to the outline of a character's shape. PostScript is the standard outline font for use in DTP.

Font output: Output quality depends on the resolution of the employed LBP as is shown in the figure: ① 16 × 16 dots; ② 24 × 24 dots; ③ 32 × 32 dots; ④ 100 × 100 dots; ⑤ phototypeset font. Aliaising of inclined strokes is improved by higher resolution. An LBP with a resolution of 100 dots or more is suitable for the production of sufficiently clear letter forms for final drawings.

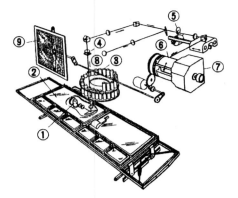

Fig. 4.219 First generation machine

Fig. 4.223 Phototypesetting fonts

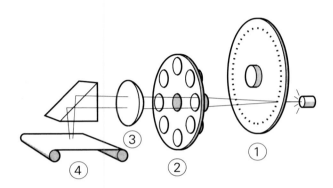

Fig. 4.220 Second generation machine

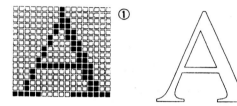

Fig. 4.224 Digital fonts

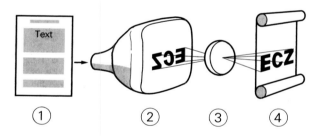

Fig. 4.221 Third generation machine

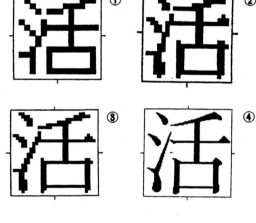

Fig. 4.225 Font output

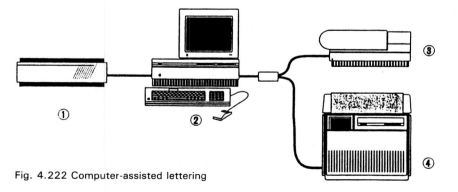

Fig. 4.222 Computer-assisted lettering

173

4.8 Aspects of computer-assisted cartography

Currently, cartography is in the midst of something of a revolution which is being caused by the ever-increasing application of computer-assisted map preparation methods. Computer-assisted typesetting is one example of this and coupled with the availability of a geographical names database provides an opportunity for the quick and easy revision of labelling whenever required. The operation can be carried out, using a personal computer, without any real need for skill or even experience in map drawing. However, successful implementation of any system allowing the computerised production of final drawings, suitable for use in the subsequent manufacture of printing plates, requires the use of sophisticated and expensive equipment, computer literacy, and cartographic expertise. Relevant technology is still being rapidly developed.

Graphics data for computer processing exhibits two distinct forms: vector data is comprised of an assembly of lines, and raster data of points. The former is captured by using a digitiser or tablet, and the latter with an image reader such as a scanner or video camera. Conversion of data from one mode to the other is not yet possible automatically, and this frequently causes problems which have to be solved by interactive, manual means.

4.8.1 Computer-assisted mask making

Map areas consisting of uniform tints or patterns can be prepared by using computer-to-final-drawing technology. Their boundaries must be digitised from a compilation to form raster data which is supplemented by the necessary attributes which are held as part of a database. Output is produced by generating and exposing each pixel on film with the aid of a high resolution LBP.

The dither method has been developed for the purpose of preparing shading with the aid of an LBP. It allows the derivation of 17 tones, between white and black, which are comprised of variations of pixel patterns rather than the normal dots of

different sizes. Each pixel is subdivided into 4×4 squares, the number of which are left blank or infilled, being varied to create a graded shading scale. There are several methods of doing this, as is illustrated in the figure: (1) Bayer's patterning method; (2) the fatten type; (3) the vertical technique; (4) the directional system.

4.8.2 Computer-assisted relief representation — 3D depiction

4.8.2.1 Computer-assisted representation of the landscape in 3D

P. Yoeli published details of a technique he called analytical hill-shading in *Kartographische Nachrichten*, **15** (4) (1965) and **16** (1) (1966). This involves the division of the area to be shaded into small, rectangular segments (pixels). The elevation of each corner of every pixel is calculated from contours by using a cubic interpolation technique developed by Lagrange. Their average inclination and orientation are computed from the relevant height data. The degree of darkness exhibited by pixels constituting the final display is determined with reference to the source of illumination. If the area of a pixel is sufficiently small, a strictly analytical shading could be generated by the computer-to-final-drawing method.

This technique, currently called automated hill-shading or computer-assisted relief representation, requires the use of height data in the form of a Digital Elevation Model (DEM) or a Digital Terrain Model (DTM). These normally consist of square arrays of spot heights, data relating to which are captured by making measurements from a contour map or aerial photographs.

In 1978 S. Murai and R. Tateishi published a paper (in *Production Researches*, **30** (7)) describing a method for the subdivision of an existing DEM to enable acquisition of a closely spaced grid of spot heights. These are interpreted, on respective subdivided grids, from approximately 16 lattice points in the DEM. This requires the use of a recently developed cubic interpolation formula based on tangent vectors on a generated cubic surface.

The latest results have appeared in an article written by K. Kobayashi in the *Journal of the*

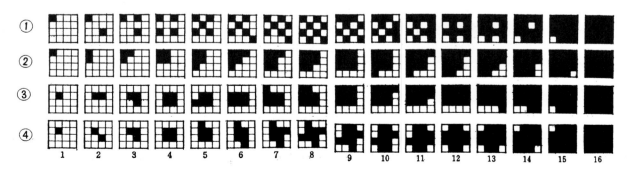

Fig. 4.226 Computer-assisted mask production. Dither patterns

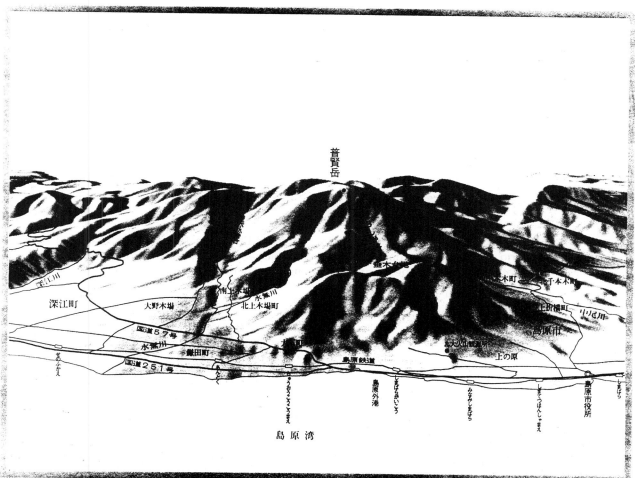

Fig. 4.227 Computer-assisted relief representation. 3D depiction

Association of Precise Survey and Applied Technology (APA), **54** (1992). The output pixel size is 0.025 mm, and the direction of the source of illumination and position of the viewpoint can be modified. This allows the production of not only various oblique shadings, but also three-dimensional (3D) depictions providing a 'bird's-eye' view, or an anaglyph derived from a DEM. The different types of output are illustrated in the figure: ① oblique shading; ② bird's-eye view of the same area.

Satellite image data linked to a DEM can be processed to produce a similar type of 3D image. With 3D depiction it is necessary to employ a hidden line removal treatment in order to delete detail which would be obscured from the viewpoint chosen.

Automated hill-shading can be produced quickly and easily, and is relatively inexpensive. Considerable detail can be represented, and connection made with information on adjacent sheets, more easily than is the case with traditional, manual drawing. However, this is only possible if an appropriately scaled DEM of the area is available and suitable processing facilities can be utilised.

4.8.2.2 Murai and Tateishi's interpolation formula

In Fig. 4.228 ① lattice points are indicated by

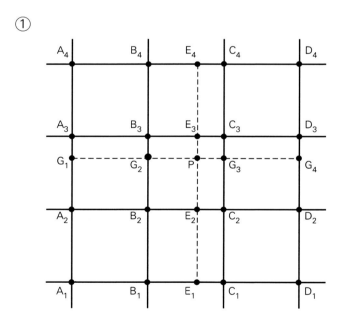

①

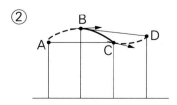

Fig. 4.228

A_i, B_i, C_i and D_i (i = 1, 2, 3, 4) on a square DEM grid. P is a point on a subdividing grid. Each grid has normalised co-ordinates — for example the co-ordinates of B_2 are (0, 0) and C_3 are (1, 1); abscissae of P and E_i are x; ordinates of P and G_i are y. Supposing that a cubic curve passes through B and C and that its tangents at B and C, respectively, are parallel to the chords AC and BD (as shown in ②), the spot height h (x, y) of P is calculated from the formula:

$$g_1 (\mu) = -(1/2) \mu^3 + \mu^2 - (1/2) \mu,$$
$$g_2 (\mu) = (2/3) \mu^3 - (5/2) \mu^2 + 1$$
$$g_3 (\mu) = -(3/2) \mu^3 + 2 \mu^2 + (1/2) \mu,$$
$$g_4 (\mu) = (1/2) \mu^3 - (1/2) \mu^2$$
$$h (x, y) = \Sigma (i = 1, 4) [g_1 (y) za_i g_i (x) + g_2 (y) zb_i g_i (x) + g_3 (y) zc_i g_i (x) + g_4 (y) zd_i g_i (x)]$$

where $g_i (x) = g_i (\mu = x)$, $g_i (y) = g_i (\mu = y)$; za_i, zb_i, zc_i and zd_i are spot heights of A_i, B_i, C_i, and D_i respectively.

For depiction in 3D, θ is the declination of the angle of view, h is the spot height of a pixel, and an originate of each pixel is converted from y to y' by $y' = y \sin \theta + h \cos \theta$.

To apply hidden line removal treatment, ($x_i \times y_j$) is the co-ordinate of respective pixels (i, j = 1, n), and j is also the order, i.e. j = 1 is the nearest to the viewing point. y'j is converted from yj. For respective i's, suppose $y'_{Max} = y'$ 1 first. Compare y'_{Max} and y'j = 1 and y'j + 1 from j + 1 = 2 to the n of the succession; if y'j + 1 ≥ y'_{Max} then $y'_{Max} = $ y'j + 1 and the pixel is left; in the converse case the pixel would be removed.

References

Aoki, I., 1972. *Cartographic Lettering Handbook.*

Bosse, H., *et al.*, 1975. *Schichtgravur.*

The College of Construction, *Textbook of Tools and Materials.*

Cuenin, R., 1973. *Cartographie Générale.*

Curran, J. P., *et al.*, 1988. *Compendium of Cartographic Techniques.*

Hodgkiss, A. G., 1972. *Maps for Books and Theses.*

ICA Standing Commission on Map Production Technology, 1991. *Flow Diagram Construction.*

Imhof, E., 1969. *Kartographische Gelandedarstellung.*

Japanese Surveyors Association, *Practical Training Text for Map Drawing.*

Kanazawa, K. & Nishimura, K., 1961. *Topographic Surveying and Map Compilation.*

Kanazawa, K., *et al.*, 1991. *Applied Cartography.*

Keates, J. S., 1989. *Cartographic Design and Production*, 2nd ed.

Krenneva, A. M., 1972. *Kartograficheskoe Cherehenie.*

Kudo, T., 1986. *Exact Usage of Drawing Tools*.

Ritchie, W., Wood, M., Wright, R., & Tait, D., 1988. *Surveying and Mapping for Field Scientists*.

Robinson, A. H., *et al*., 1993. *Elements of Cartography*, 6th ed.

U.S. War Department. *Topographic Drafting* (TM5−230).

Chapter 5

CARTOGRAPHIC PRE-PRESS, PRESS AND POST-PRESS PRODUCTION

C. Palm and S. van der Steen

CONTENTS

5.1 Introduction

Modern maps may be produced by traditional methods, with assistance from computers or, as is becoming increasingly common, by a combination of both conventional and automated techniques. In all cases it is essential that the cartographer is fully familiar with both the complete mapmaking processes and the materials necessarily employed during their operation.

This chapter, which considers 'Cartographic Pre-Press, Press and Post-Press Production', will deal with these techniques and materials, but will also emphasise the value of traditional methods which can never be completely ignored during multi-copy generation.

5.2 Employed terminology

Analogue data — detail relating to representation by means of continuously variable physical quantities.

Assembling — the bringing together or mounting/montage and imposing of materials such as maps, films, etc.

Base — a support medium for emulsions from which images are formed.

Binary — a counting system based on the figures 'O' and '1'. Sometimes shortened to 'bit', the abbreviated form of the digital information measurement unit the 'binary digit'.

CCD scanner — an instrument which uses sensors to collect light reflected or transmitted from original documents such as maps or photographs. The acronym 'CCD' stands for 'Charge Coupled Device'.

Colour proof — an interim document, produced prior to final printing, enabling the checking of an intended map's content in terms of its appearance, accuracy, quality, colour and registration.

Colour separation — the conversion of a multi-coloured original, with a colour filter interposed, into monochrome film in order to produce a negative from which the filtered colour has been eliminated. This can be accomplished by using a process camera or optical scanner, and normally results in production of cyan, magenta, yellow and black separations.

Contact box — an instrument used for the same-size generation of either positive or negative films by exposure of an original, with a new piece of film, to white or ultraviolet light. Contact is maintained by a vacuum pressure system.

Contact frame — similar to a contact box, but larger and with the source of illumination separate from the glass-topped frame which holds materials under vacuum pressure.

Contact photography — a same-size, photographic process involving changing an image from positive to negative or vice-versa; the copying/duplication of positive or negative images; modification of image position; combination of images;

180

	conversion of transparent continuous-tone images to halftones; the production of halftone tints.	Digitising	(numbers). — the conversion of an analogue data representation to a digital one; thus representing a position on a surface by a pair of finite co-ordinates.
Continuous-tone	— a term relating to the appearance of an image which consists of various shades/differences in tonal value ranging from black to white.	Duplication	— the reproducing of an identical copy from positive or negative film. Also called a direct positive, reversal or auto-reversal technique.
Control blocks	— elements reproduced on film or printed materials in order to assist in the control of printing quality. They often consist of density patches, resolution elements, screen check devices and ink-water test images.	Emulsion	— a light-sensitive, gelatine coating applied to a film or paper base. It contains chemicals such as diazo, silver halides or photopolymers.
Darkroom	— an area dedicated to the controlled processing of photographic materials in order to produce negatives or positives for employment in the production of maps or other printed materials. Illumination is normally provided by red safelights, but sometimes yellow ones are also available.	Finishing	— the process, or processes, undertaken after printing and relating to the form and appearance of a cartographic end-product.
Daylight	— conditions experienced when all of the constituents of the visible spectrum, together with some ultraviolet radiation, are present.	Fixing	— an integral part of film processing which stabilises the chemicals composing the image areas which remain after development and the washing away of unwanted silver halides.
Densitometer	— a calibrated instrument for the measurement of the density of both transparent and opaque materials such as film, photographic and printing papers.	Folding	— a post-press production operation resulting in the generation of a folded map or multi-page atlas section.
Density	— a logarithmic standard relating to the amount of non-reflected or non-transmitted illumination.	Gum arabic	— a liquid gelatine applied to the surface of a printing plate to prevent the corrosion, and enhance the water receptivity, of non-image areas.
Density range	— the subtraction of the minimum from the maximum density of an image.	Halftone	— a photographic technique whereby a solid image is broken up, by using a screen, into evenly spaced dots of varying sizes. It gives an impression of continuous tone.
Development	— the production of a visible from an invisible, or latent, image formed by exposure. This is accomplished by means of a chemical or physical process.	Intaglio	— a printing process in which the image is recessed in the printing surface.
Diazo	— an ultraviolet light-sensitive chemical compound used to produce cheap blueprints, guideline images, colour proofs and printing surfaces.	Lamination	— the application of a thin transparent plastic sheet, usually with a high gloss finish, to the surface of a map in order to improve its appearance and increase its durability.
Digital	— pertaining to computer stored data in the form of digits	Letterpress	— a 'relief' printing process in which the image areas stand above the supporting surface.
		Line photography	— a facet of photography which relates, primarily, to the reproducing of line and text elements.

Linework — an image consisting of discrete elements, most of which are linear in character.

Lithography — a 'planographic' (surface) process which is the most commonly employed method of printing maps. The principle of grease not mixing with water during the alternate damping and inking of a plate is the key to this technique. 'Offset lithography' is a development of this, and involves indirect printing with the image being transferred from the plate onto a rubber blanket, and then from here to the paper.

Masking — the superimposition of opaque material on a positive or negative in order to exclude light from predetermined areas during photomechanical working. The resultant 'open windows' can then be used in association with screens to produce area symbols.

Merging — the combination of detail. The term is mainly used in computer cartography and refers to operation on two or more ordered sets of records to create a single set in one file.

Metal halide — a type of gas used in lamps which produce ultraviolet illumination.

Negative plate — an offset printing surface normally consisting of an aluminium base covered with an ultraviolet sensitive emulsion which hardens when exposed. It is only used with wrong-reading negative originals.

Offset plates — aluminium, or occasionally zinc, plates used in the generation of images by offset lithography.

Offset press — a printing machine which reproduces images using the offset method involving the transfer of detail from a plate onto a rubber blanket and then to paper.

Orthochromatic — a photographic term which refers to sensitivity, and relates specifically to the white-light spectrum with the exception of red.

Pagemaking — the assembly of film elements constituting a page, or series of pages, in an atlas.

Peel-coat — an ultraviolet light-sensitive masking film used in the production of either negative or positive masks during map preparation by traditional means (see also Strip mask).

Photocopy — a facsimile version of an original document produced on paper by electrostatic principles.

Photo-polymer — a recently developed, light-sensitive material forming an integral part of some pre-press proofing systems. It may be used as a hardening agent, for making things soluble, or removing stickiness.

Plastification — the applying of plastic coatings to paper or board.

Plotter — a computer-controlled device capable of generating permanent graphic images from digital data. Lines, text, symbols, patterns and tones can be produced. The accuracy of output depends on the type of plotter and its purchase price.

Polyester — the most frequently used, transparent, dimensionally stable base material employed in map reproduction.

Positive plate — an offset printing surface on which the ultraviolet-sensitive emulsion forms the image area when exposed with reversed positives. The pre-sensitised layer is composed of a diazo or photo-polymer substance. If the emulsion is manually applied, it consists of dichromated colloids.

Pre-press, Press, Post-press — descriptive expressions employed in the contemporary mapping and graphics industries, and relating to the generation of images, their printing, and finishing.

Process camera — a precision instrument, with a large negative format, for the copying (including reduction and enlargement) of drawings and other graphic documents intended for eventual reproduction in print.

Proofing press	— a flat-bed, offset printing press which is operated either by hand or mechanically. It is used to produce colour proofs and for small runs.
Punch-registration	— a method whereby a series of precisely located holes are mechanically created in the margins of either reprographic materials (films, etc.) or printing plates prior to their use. Studs or a pin-bar are inserted in the perforations of the first sheet, and other components are overlaid in such a way that they pass through corresponding holes, so ensuring perfect register or 'fit'.
Rapid access	— an expression applied to the quick processing of photographic material as a result of increasing the temperature of both the employed developer and fixer. This technique can only be used for linework, and requires purchase of a special type of film.
Register marks	— appropriate crosses or lines incorporated on all drawings or films to facilitate the preservation of scale and register at all stages of production.
Resolution	— a term referring to the grading of the ability of film to incorporate the maximum amount of fine detail. It is expressed as the number of lines visible per mm/cm/inch.
Re-touching	— this can apply to either the manual correction or improvement of photographically generated images; or the editing/modification of digital images stored in a computer and called up for screen display.
Safelight	— a source of illumination used in pre-press processing rooms which does not affect light-sensitive emulsions.
Scanner	— an instrument which reproduces images by scanning them with sensors. This implies the duplication of the density of amounts of reflected light, and storing this in a computer in digital form.
Screens	— these consist of sheets of transparent film incorporating lines, dots, or other regularly repeated patterns which may be used in conjunction with an 'open-window' mask, either photographically or photomechanically, to reproduce areas of pattern. Contact/film screens on a flexible base (whose transparent elements are graded in density) are used to convert continuous-tones to screen halftones.
Scribecoat	— a photographically opaque material consisting of a coating, on a translucent polyester film base, through which transparent lines or other symbols can be engraved by using scribing tools.
Sensor	— a component part of a scanner which transforms transmitted or reflected light into electronic signals.
Silk-screen	— a relatively cheap printing process which allows creation of images by using a fabric composed of polyester threads. Ink can be squeezed through this onto a printing surface.
Silver halides	— light-sensitive components of photographic emulsions which serve to distinguish the latter from other coatings. They can be used in conjunction with a wide range of light sources.
Skeletonising	— the reduction of raster format data relating to linear features to the width of a single pixel. This serves to represent the centre of the original line.
Spectrum	— the visible part of the electromagnetic spectrum lying between, approximately, 400 and 700 nm.
Strip film	— a photographic film which allows removal of a thin emulsion membrane from its supporting base after exposure and processing. Subsequently, this, or elements from it, can be remounted and positioned on other material.
Strip mask	— a polyester-based material which is overlaid with a thin, photographically opaque, red,

orange, or yellow membrane. The latter is easily cut or etched, and is subsequently removed to generate either a positive or negative mask (see also Peel-coat).

Substratum — this general term is used to indicate the type of base material employed in the manufacture of various reprographic products.

Test strips/ step wedges/ control blocks — these photographic or printed elements are positioned outside the trim marks indicating intended format, and serve as a quality control index during photomechanical working or printing.

Trim marks — lines or corner marks reproduced, on films or printed copies, to indicate required format and consequently the amount of material to be removed during trimming.

Trimming — the finishing process involving the guillotining/cutting away of excess paper around the edges of a printed map in order to achieve the desired format.

Ultraviolet (UV) — the element of the electromagnetic spectrum lying, approximately, between 300 and 425 nm. This type of illumination is used in association with a variety of reprographic processes, and may also perform an important rôle as part of special drying systems.

Vacuum pressure — this is regularly used in association with a contact box or frame. It is necessary for the maintenance of perfect contact between a working document and sensitised materials of different types.

Varnishing — a finishing process sometimes adopted to give a more appealing appearance, whilst also serving to make a map printed on paper more durable and weatherproof. Varnish may be applied manually or by a printing press.

Vectorisation — a process resulting in the transformation of raster data into a vectorised form.

Whirler — a machine consisting of a revolving bed on which substrata, or aluminium/zinc printing plates, are positioned in order to receive a coating with a liquid such as dichromated colloid, colour dye, etc. The spinning of the bed ensures even distribution of the coating as a result of the action of centrifugal force. Although not used as regularly as was formerly the case, whirlers are still employed by some cartographic organisations for both proofing and platemaking.

5.3 Pre-press production

5.3.1 Darkroom processes

5.3.1.1 Films and papers

The photographic emulsions coated on materials used in a darkroom contain 'silver halides' which are sensitive only to the 'non-red' parts of the visible spectrum. These orthochromatic emulsions are sensitive to white light, and can thus only be used under red/safelights in a darkroom environment. Continuous tone, halftone, and in some instances line photography are carried out under these conditions of illumination. However, in more modern establishments 'contact linework photography' is now performed under yellow/daylight working conditions and involves the application of ultraviolet (UV) processes. The camera is still necessarily operated in a darkroom, and can be used to produce continuous tone, halftone and linework.

5.3.1.2 Processing

The pre-press cartographic processing of these emulsions requires use of a film and paper processor containing a special developer for continuous tone work; an alternative one for linework (rapid access) processing; and lith chemistry for halftone processes (Fig. 5.1).

Manual developing must take place in a red-light environment, and involves the use of specific solutions to process photographic images. Increasingly, these methods are being replaced by the automatic development of photographic materials as a part of the ongoing sophistication of cartographic reproduction.

5.3.1.3 Process camera photography

Cameras play an important role in the reproduction of original, analogue data; for example, maps, aerial photographs, text and other forms of illustrations. Two main types are manufactured.

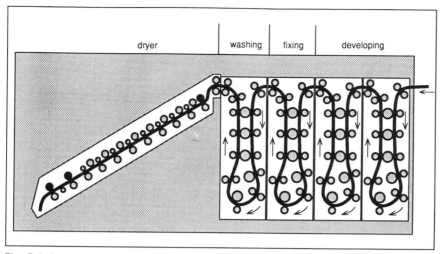

Fig. 5.1 Automatic processing of photographic materials

Firstly, the 'horizontal' (two-room) version in which the camera back is built into a darkroom wall (Fig. 5.2), and secondly, the increasingly popular 'vertical' camera (Fig. 5.3).

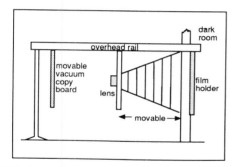

Fig. 5.2 'Horizontal' (two-room) camera

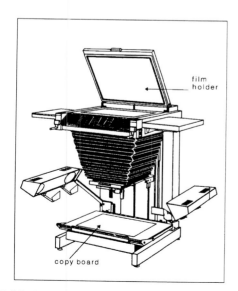

Fig. 5.3 'Vertical' camera

The latter is considerably cheaper than the former, but its potential reproduction size is more limited—a factor which should be recognised before placing an order!

In traditional cartographic reproduction the camera performs the following functions: the same-size reproduction of opaque materials; the enlargement of either transparent or opaque originals; the reduction of either transparent or opaque originals; the rectification of detail. This is only possible if the instrument is equipped with a tilting copyboard.

5.3.1.4 Contact photography

'Contact frames', and the more frequently employed 'contact boxes', are used in a darkroom to prepare same-size reversals of materials. Transparent or translucent originals are positioned in contact with sensitive photographic film or paper and, under vacuum, are exposed to a tungsten 'white light' or 'ultraviolet' metal halide light source. The former can also be used, in combination with coloured gelatine filters, for special purposes such as colour separation (Fig. 5.4).

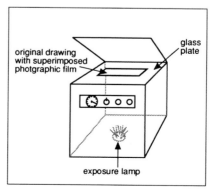

Fig. 5.4 'Contact box'

It must be stressed that 'contact photography' allows only the 'same-size' reproduction of originals, and that if enlargement or reduction is required a process camera must be employed. Red-lighting in a darkroom must not be too bright because of the ultra-sensitive nature of orthochromatic film.

Contact photography can be used for: reproducing transparencies/films; combining continuous-tone films with text and/or linework; producing so-called 'fat' (thickened) stop-out masks; changing continuous-tone images to halftone products; generating right-reading and wrong-reading results (see Fig. 5.5).

In the event that contact equipment is not available in association with UV facilities, this type of photography can also be used for: producing negatives from positive films; positive to positive duplication; the combination of negatives, masks, etc., to generate final positives (see Figs 5.6 and 5.7); the production of stripping film for type, symbols, etc.

5.3.1.5 Corrections

As with all production processes, it is essential to be able to carry out the correction of unforeseen errors which appear on final materials. It is, of course, possible to prepare an amended version of a camera- or contact-exposure, but it is often quicker and cheaper to make corrections. These can be carried out in the following ways:

—the removal of 'white spots' by duffing a negative with either brown or black opaquing fluid;

—the removal of 'black spots' by either scratching on the emulsion side of a film positive with a knife or applying a liquid remover;

—the replacement of elements of type or linework by 'cutting out' erroneous detail and inserting correct material;

—the addition of linework by drawing it on matt drafting film;

—the addition of type by using adhesive or waxed stripping film.

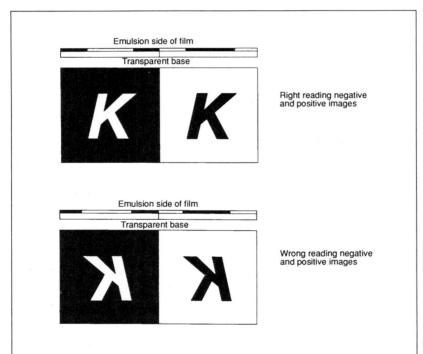

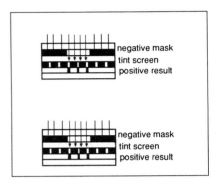

Fig. 5.6 Contact copying involving combination of a negative mask and a tint screen to produce a 'same-size' final positive

Fig. 5.5 Contact photography to produce 'right' and 'wrong' reading results

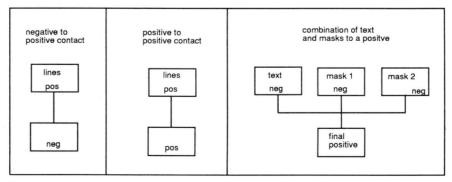

Fig. 5.7 Various methods of contact copying and detail combination

5.3.2 Ultraviolet (UV) processes

Processes employing UV-sensitive emulsions have been evolved for use in 'diazo printing', 'mask preparation', 'colour proofing', 'platemaking' and the application of guide images to drafting film and scribecoat.

Since the late 1970s it has also been possible to purchase contact photographic films and papers, for linework, with UV-sensitive coatings. This means that difficult and critical accuracy jobs no longer have to be carried out under red-light darkroom conditions, but rather that they can be performed using the same yellow-light/daylight working conditions applicable to other processes. If the photographic facilities within the pre-press section of a cartographic organisation are large, specific yellow-light rooms may be constructed for UV contact photography.

Unfortunately, it is not possible to use UV light-sensitive emulsions in association with a process camera. This is because they require a very intense exposure source and, as lenses permit the passage of only a small quantity of light, the strength of the rays reaching the film is insufficient to create a reaction with the UV-sensitive photographic emulsion (Fig. 5.8).

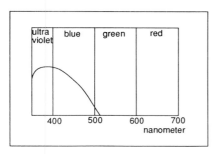

Fig. 5.8 Light sensitive curve for UV photo emulsions

5.3.2.1 Contact photography
UV contact frames and boxes constitute the hardware necessary for the production of either positive or negative results from linework and/or halftone transparencies. Films available for this type of photography are slow and require exposure to a very intense UV light source. The most commonly employed illuminant is the Metal Halide Lamp which is equipped with air-cooling to control bulb temperature.

5.3.2.2 Films and papers
As with 'darkroom films', a variety of UV-sensitive emulsions are available. For example: films and papers allowing contact processing from negative to positive or positive to negative; duplicating film for use in contacting from positive to positive or negative to negative.

The practical processes relating to contact reproduction are executed in the same way as conventional contact copying in a traditional darkroom. Appropriate contact between original documents and photographic materials is ensured by employment of a vacuum system, whilst the measured time exposure of the sensitive emulsions to a controlled strength UV light source is determined by pre-set photo-electric cells within the instrument.

5.3.2.3 Processing
In principle, there is no difference between the processing of darkroom films and papers and that of yellow-room/daylight working photographic materials. The 'development' of images is followed by their 'fixing', 'rinsing' in water, and subsequent 'drying' using specific equipment. All of these operations are carried out in a 'light-safe' processor of the type used in a normal darkroom. Because of this, processing equipment is often located in areas which can be either red- or yellow-light illuminated.

UV emulsions can also be processed by manual methods, but once a change has been made to use of these materials, traditional techniques are normally replaced by automated equipment.

5.3.2.4 Corrections
See paragraph 5.3.1.5 relating to darkroom processes.

5.3.3 Diazo

Diazo coatings comprise one of the earliest developed series of light-sensitive chemicals. For many years they were employed in the production of cheap 'blue prints/blue lines', either on paper or transparent materials, which were used as guide images in the duplication of detail relating to linework or text.

During the 1970s the potential afforded by diazo chemistry was rediscovered with reference to the manufacture of offset plates. However, the production of blue prints constituted the principal use of the process in the interim. Diazo chemicals contain components which are activated when brought into contact with ammonia fumes. They result in the production of a coloured dye without the surface being exposed to UV illumination. When, even if by accident, sufficient light has affected the emulsion, the diazo coating will be broken down into its constituent chemicals and image formation is impossible.

Nowadays diazo is employed in the manufacture of the following: diazo 'paper' (red, blue, brown and black); diazo 'film' (both clear and matt for red, brown and black images); diazo 'colour film' (clear base only); pre-sensitised 'printing plates'; 'guide images' printed down onto drafting film or scribecoat.

The last of these can also be produced 'in house' by the manual application of a 'rub on' liquid. This can be purchased in a pre-prepared form or as two chemical powders and an appropriate liquid which are mixed prior to use.

5.3.3.1 Processing
The 'processing' of diazo materials is dependent on the way in which the product is prepared. The majority of them consist of pre-sensitised chemicals requiring ammonia development. This means that, subsequent to UV exposure, the diazo chemicals must necessarily be subjected to ammonia fumes in a special processor (Fig. 5.9).

5.3.3.2 Chemicals
Diazo 'chemicals' used in the manufacture of printing plates serve to create image areas and therefore also contain other chemicals. The image held on a plate is at a slightly higher level than non-image areas. The diazo constituents of the sensitised coating do not require exposure to ammonia fumes, but are developed using an especially prepared liquid (see also subsection 5.3.1).

5.3.4 Mask production

Essentially a mask is an original document prepared by the cartographer for use in producing area images comprised of solid colours, screened tints or patterns. However, it can also be employed to simplify or block out certain details contained within an original.

It can be generated in a positive or negative form (Fig. 5.10), and consists of a clear polyester base (foil) on which the areas intended to prevent light transmission appear in an opaque form. The degree to which precise registration of detail is necessary determines the masking technique adopted.

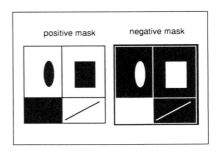

Fig. 5.10 Mask production

5.3.4.1 Simple method
The most straightforward application of this method involves drawing on a transparent sheet with a special type of ink (not that normally used for drawing), water-soluble opaquing fluid, or a retouching pen. Alternatively, yellow or orange lithographic paper can be mounted onto the areas concerned. These techniques can be used singly or in combination.

5.3.4.2 'Cut and peel' method
Commercially produced materials are available, either as sheets or in roll form, consisting of a very thin, peelable, red membrane which adheres to a clear, polyester base/carrier sheet. It is carefully superimposed, with respect to appropriate fit marks or by using a punch-registration system, on a positive original which outlines the desired areas. Subsequently, a scalpel or swivel-bladed knife (specially designed for the purpose) is used to cut around the outlines of the areas concerned, and extraneous materials are peeled away. Either positive or negative masks can be prepared in this way. The process is illustrated in Fig. 5.11.

5.3.4.3 Strip mask or peel-coat
This rather more expensive material is used when precise registration of detail is important. Open lines relating to intended detail are photomechanic-

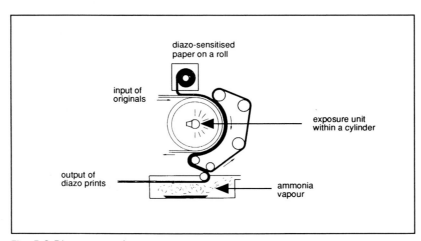

Fig. 5.9 Diazo processing

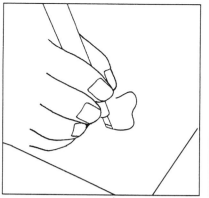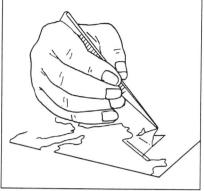

Fig. 5.11 Mask production by the 'cut and peel' method

ally produced on the opaque membrane. The polyester base material is the same as that used for 'cut and peel', and the overlying red or orange film is pre-sensitised for either positive or negative working. An original 'same-size' document is exposed, in a vacuum frame, in contact with the sensitised layer of the peel-coat. Use of a 'right-reading (RR)' original results in generation of a 'wrong-reading (WR)' peel-coat. After development, the exposed linework is etched to reveal the clear base material (Fig. 5.12). The sheet is then ready for peeling.

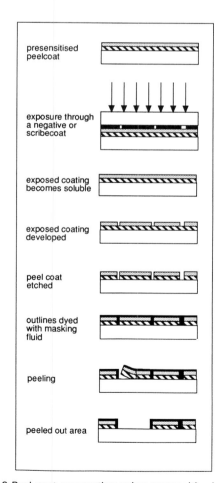

presensitised peelcoat	
exposure through a negative or scribecoat	
exposed coating becomes soluble	
exposed coating developed	
peel coat etched	
outlines dyed with masking fluid	
peeling	
peeled out area	

Fig. 5.12 Peel-coat preparation using presensitised, negative working material

All of the intended map elements appear as clear detail on a strip mask, and linework will remain in some areas after peeling has taken place. However, it is possible to opaque out unwanted lines before peeling by applying dye to the complete sheet, or the relevant part of it, using special masking fluid. Alternatively, normal, water-soluble, opaquing medium or a retouching pen can be used for this purpose.

It is also possible to purchase sheets of opaque, adhesive-backed, red material which can be used to construct small or simple masks, ideally of a regular shape (legend boxes, etc.). Lithographic tape can be used for the same purpose.

Masks produced by peeling are very fragile and easily damaged. Consequently, they must be handled extremely carefully or, if they are to be stored for further use, photographically copied.

5.3.5 Colour proof production

5.3.5.1 Introduction
The principal functions of a colour proof produced by a cartographic organisation relate to the evaluation of map content in terms of its accuracy, quality and colour, and to ensuring that the various reproduction components are complete and appropriately registered, one with another, prior to the final printing of the document. In general terms, the main groups of proofing techniques can be summarised as: press proofing with inks; photomechanical proofing.

The former is ideal if the proof is to be of a prototype map required as a guide to colour fidelity, or for approval by a customer. However, this method necessitates the manufacture of printing plates, one for each colour, and the use of either a special offset proofing press or regular production equipment. This is both expensive and time-consuming and, as a result, proofing with inks is increasingly rarely used in cartographic establishments. The remainder of this section will therefore concentrate on methods from the second category and discuss photomechanical systems which are capable of producing proofs directly from reproduction materials which normally consist of combined positives or negatives for each of the desired colours.

Proofs are frequently required at various stages of map production, and their functions are much broader than merely colour control. It is therefore necessary to make a clear distinction between a proof *of* colour and proofing *in* colour. In practice, where the accurate registration of the assembled map components is of greater importance than colour fidelity, the most commonly used methods belong to the latter group, i.e. they are used to create proofs *in* colour.

5.3.5.2 The classification of photomechanical colour proofing systems

A comparison between different proofing systems involves several factors, the most important of which are:

- **Physical type**: The main distinction here is between 'superimposure' (single sheet) methods in which all colours are printed down onto the same piece of base material; and 'overlay' proofs where each of the colours constitutes a separate transparent layer. The latter, although often satisfactory as proofs *in* colour, are not appropriate for making proofs *of* colour because of the reflections produced by the various layers of translucent material.

- **Range of colours**: Many of the commercially available graphic arts proofing methods relate to multi-colour reproduction using the standard four-colour process (cyan, magenta, yellow and black). Others provide only a limited range of colours which it is possible to combine or mix. This facility, enabling the creation of a particularly required colour, is often of significant advantage in cartography.

- **Format**: Commonly, map production results in the generation of large format imagery. A number of potentially useful systems may prove of limited value because the available working area is too small for many mapping requirements.

- **Substrata** (base materials): If proofs *of* colour are required, it is virtually essential to produce these on the same paper stock that it is intended to use during final printing. For proofs *in* colour, particularly in the case of the 'overlay' system, it is necessary to employ a soluble coating and liquid dyes in association with a polyester or polyvinyl base material.

- **Reprographic material**: All pre-press photomechanical proofing methods require exposure, through either a combined positive or negative, to bring about image transfer. The difference involved in working with either positives or negatives is important. For example, in most cartographic production methods, map components are likely to be in negative form, in which case access to a negative proofing system is obviously important. Conversely, if the elements are in a positive state then use of a positive working system is highly desirable. The need to generate proofs, from either positives or negatives, is one reason why many cartographic agencies are reluctant to invest in technology which can operate from one, rather than both, types of reprographic materials. The fact has now been accepted by equipment/material manufacturers, and the most up-to-date products allow either positive or negative processing.

- **Special equipment**: The majority of production agencies have access to large format exposure and developing facilities, together with contact frames, sinks and possibly a drier. Some may still have a 'whirler' which can be used to coat plastic sheets, but these are now much less common than was formerly the case. Both 'superimposure' and 'overlay' systems have been developed which do not require use of any specialised equipment, and thus only the light-sensitive coatings, dyes or pre-sensitised materials have to be purchased to allow proof production. Other techniques require special processing machinery. Investment in this has to be considered with regard to the total amount of proof work likely to be undertaken, and also whether the hardware is capable of the type and size of proof wanted.

- **Cost**: The price of producing a proof depends on the equipment and materials used, and also the man-hours involved in processing. It is also affected by the skill levels of responsible personnel and the total amount of proofing undertaken. If large numbers of standard format and styled proofs are required, the high purchase costs of necessarily expensive machinery may soon be justified. However, if relatively small quantities are needed, and have to be made from a wide variety of negative and positive elements, such an investment may be rather harder to justify! Often it is possible to lease machinery, and this is a factor worthy of consideration.

5.3.5.3 Methods

The different systems manufactured can be divided into the following groups, depending on the methods employed in their operation. Some examples of commonly used and readily available commercial brands are cited:

- rub-on/wipe-on superimposure processes ('Kwik-Proof');
- electrostatic proofing (Kimofax and Signature systems);
- transfer methods involving powders or the laminating of coloured films (Cromalin (powder), Matchprint or Agfaproof (film));

—overlay systems (Color-Key, Celsia Proof, Naps/Paps);
—coating bars (Coloreae, Matro-Color).

5.3.6 Platemaking for map printing by offset lithography

The multi-copy printing of maps is normally undertaken using the lithographic process and offset presses. These machines use aluminium plates from which the image is 'offset' onto a rubber blanket and subsequently transferred to paper. Aluminium is used for plate manufacture because it is: strong and durable; light in weight; easily 'grained' using either electrical or chemical processes; sufficiently malleable to be easily bent around the plate cylinder on the printing press; attractive to water but corrosion-resistant; potentially recyclable.

5.3.6.1 Printing plate metals

Although in earlier times 'zinc' was the metal primarily used in plate manufacture, this has now been largely replaced by 'aluminium' or bimetallic and trimetallic bases. Some mapping organisations still employ a whirler to apply a light-sensitive coating to one side of the aluminium, but in most, pre-sensitised plates are now purchased for use in printing by offset lithography. The surface of the plate is grained in order to be water receptive in non-image areas. Greasy ink adheres to image areas and repels water when the plate is damped with a thin film of water, on a regular basis, during the printing process.

5.3.6.2 Light-sensitive coatings

'Pre-sensitised printing plates' consist of a metal base covered with a UV-sensitive layer which constitutes the printing surface. The light-sensitive chemical can consist of either diazo compounds or photo-polymers (see Fig. 5.13).

Formation of an image depends largely on the chemical materials used to make the coating, and both negative and positive working plates are available.

'Negative' plates carry an emulsion which is hardened if their surface is exposed to UV illumination. Non-exposed areas remain soluble and can be dissolved using an appropriate developer. These are known as 'surface plates', because the image lies entirely on the surface.

'Positive' plates react in the opposite way and UV light renders the chemical coating soluble, whilst areas of unexposed emulsion remain hard. The exposed background is dissolved and the remaining parts constitute the printing image. Again, a surface image is formed, but the positive process is generally used to produce 'deep-etch' plates on which the image is slightly recessed in the surface.

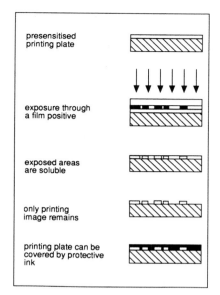

Fig. 5.13a Manufacture of positive working presensitised printing plates

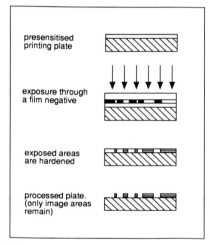

Fig. 5.13b Manufacture of negative working presensitised plates

5.3.6.3 Offset plate processing

Plate 'development' is carried out by using a pad to apply a liquid developer appropriate to the type of coating present. This is wiped over the whole surface of the plate, using a circular motion, until all of the unwanted coating has been dissolved.

To ensure correct processing, a special test-strip/step wedge is exposed on the plate at the same time as the original film, so enabling the control of development. When this is satisfactorily completed, the plate is rinsed thoroughly with water to ensure that all of the unnecessary chemicals are removed. Subsequently, 'corrections' may be made, or the plate will be covered with a protective layer of 'gum arabic' which also serves to make non-image areas more water receptive. This solution does not adhere to the

191

image areas which will later be inked, but which at this stage are uncoated.

Plate processors make development both quicker and easier, and the results obtained by using them are much more regular. Their employment is particularly recommended when significant quantities of maps are to be produced, but the quality improvement is of minor importance.

After processing, plates should be used as soon as possible, because daylight may affect the quality of the printing surface.

5.3.6.4 Plate correction
Modification of plates is sometimes necessary in order to delete deliberately included test strips or step wedges; blemishes possibly resulting from unnoticed dust or dirt which has accidentally landed on the surface prior to exposure; and also unavoidably reproduced film montage lines.

Specific correction agents, such as gels or fluids, are carefully applied to the plate surface and dissolve unwanted detail. This is often necessary in positive working, but negative plates tend to need less correction.

5.3.6.5 Recent advances in platemaking
A recent innovation is the introduction of so-called 'dry plates'. These consist of an aluminium base which supports a 'photo-polymer' and a transparent 'silicon' coating which is, in principle, an ink-repellent surface. When exposed to UV light, the photo-polymer bonds itself to the silicon. Processing by rubbing causes the remainder of the silicon to be removed, and the ink receptive polymer appears at the surface. Plates of this type are positive working.

Other developments relating to offset printing have included the introduction of bimetallic and trimetallic plates. These have a particular advantage in that they enable print runs well in excess of 200 000 copies. However, it is only very rarely that more than 100 000 copies of a map are produced, and so this type of plate is normally unnecessary.

5.3.6.6 The use of printing plates
A variety of organisations employ offset printing, and therefore offset plates, in their generation of graphic materials such as books, magazines, newspapers, advertisements and, of course, maps.

After the necessary multi-copy production has been completed, plates should be stored in a cool, dark place. Often they can be recycled by specialist companies and it is therefore sensible to collect quantities of used plates for subsequent reprocessing.

5.3.7 Silk-screen making and printing

5.3.7.1 Introduction
This is a stencil printing method which is not often used for map printing, but which can sometimes be useful in doing jobs which are impossible using the more complicated offset process. A negative stencil is formed on top of a fine mesh or fabric, which is tightly stretched over a frame, and ink is forced through this under pressure. The resultant image is comprised of tiny points or dots. Initially, the mesh was formed of silk, but today consists of either nylon or polyester threads with a density of between 48 and 132 lines per centimetre (120 and 330 lines per inch).

Having positioned the frame over the paper (or other intended printing surface), ink is pressed through the unmasked areas of the screen using a rubber squeegee blade (see Figs 5.14 and 5.15).

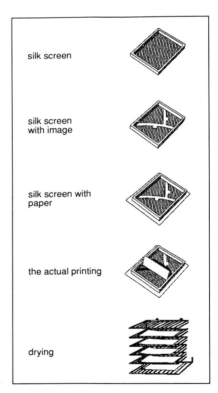

silk screen

silk screen with image

silk screen with paper

the actual printing

drying

Fig. 5.14 The process of silk-screen printing

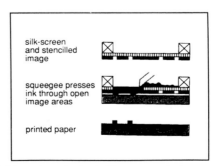

silk-screen and stencilled image

squeegee presses ink through open image areas

printed paper

Fig. 5.15 Manual method of silk-screen printing

5.3.7.2 Direct method

Preparation of the stencil may involve the manual coating of the silk-screen to produce a mask, but it is more normal that the negative graphic is photographically reproduced directly onto the fabric.

5.3.7.3 Indirect method

See Fig. 5.16.

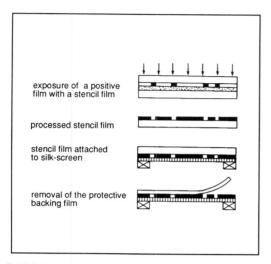

Fig. 5.16 Indirect method of silk-screen printing

5.3.7.4 Combined method

This is the most commonly used technique and involves the application of a light-sensitive photopolymer onto and into the mesh. When it has dried, it is exposed with a right-reading positive original, and is then developed in water to create the necessary open-window areas (Fig. 5.17).

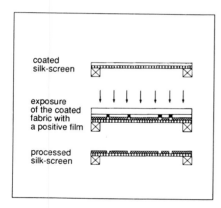

Fig. 5.17 The combined method of silk-screen printing

5.3.7.5 Silk-screen printing

The generation of copies can be performed manually or by complicated presses capable of producing between 2 000 and 10 000 impressions per hour. It is only possible to print one colour at a time because the ink is very slow to dry. The process is seldom used to produce complete maps but, as it is easy to maintain register, it can be used to solve problems such as the overprinting of detail on finished sheets which have already been trimmed.

It can also be used in the printing of line and halftone work, employing either fluorescent or opaque inks, on plastics, cotton, glass, bottles and board (of any type or thickness). A particular example of the use of this technique is provided by outdoor map displays at the roadside or in parks. These often involve the printing of light-fast inks onto plastic panels in order to maintain the strength and visibility of details even under bad weather conditions. Other applications can include wrappers, polished surfaces, overhead transparencies and printed circuits.

The major advantages of this type of printing are:
— it is easy to maintain register even when printing on trimmed map stock;
— either transparent, opaque, bronze or fluorescent inks can be used;
— although some basic training is necessary, it is a much simpler technique to learn and apply than offset lithography;
— only a relatively low financial investment is necessary, even for the purchase of an appropriate press.

However, there are disadvantages which include the facts that:
— it is time-consuming, especially when printing large numbers of copies;
— the inks employed are very slow drying;
— edition quantities are limited;
— it is very difficult to print detail manually using screens finer than 48 lines per centimetre (120 lines per inch).

5.3.8 Image registration

Cartographic production requires the accurate superimposition of a variety of different elements which may ultimately appear in one or more printed colours. Included details can relate to information represented by point, line and area symbols or text, each of which may appear on separate transparent overlays which have been prepared using dimensionally stable materials. Additionally, their effective presentation may require the employment of different colours or even the combination of these. The accurate 'registration' or 'fitting together' of the various components can be greatly assisted by the use of 'register marks' and a 'punch-registration' system.

5.3.8.1 Register marks

These are the devices most commonly used by the cartographer to ensure the accurate and appropriate fitting together of overlays containing different features. They consist of a finely drawn

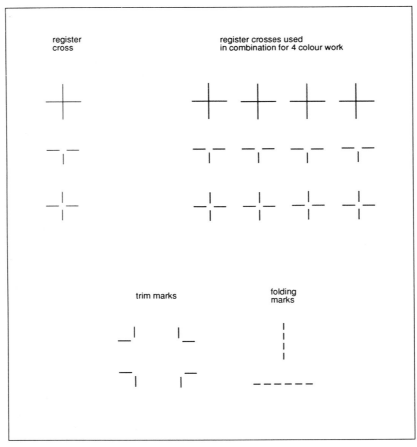

Fig. 5.18 Elements for printing folding and trimming paper for maps and atlases

cross (or crosses) or refinements of this symbol. In addition, trim marks can serve a similar purpose (Fig. 5.18).

Register marks are always positioned just outside the area occupied by map detail in order that they will be trimmed off at a later stage and not appear on the finished product. At least two of them should be included on each overlay, and they are normally located at opposite sides of the map and approximately in the middle of each of the chosen sides. They can be constructed in a variety of ways: by drawing in the intended position; by scribing, subsequent multiplication on photographic film and mounting on the overlays; reproduction from so-called 'mother crosses' incorporated on a page master; their specification from amongst existing symbols held as elements within a cartographic software program (see subsection 5.3.10).

Crosses are precisely positioned on each new overlay with reference to their original placement on the base document. Normally they are drawn or mounted in position prior to the drafting of any other details. However, sometimes they are generated in the darkroom and exposed onto film or paper with other map constituents.

Their inclusion is an essential and indispensable aid to a printer in his eventual assembly of different coloured elements.

5.3.8.2 Punch-registration

Although useful during the drafting stages, the employment of register marks can cause problems in a darkroom or under subdued lighting conditions when they tend to be very difficult to see!

To overcome problems of this type, and also to make registration easier during daylight working, a kind of perforating device has been developed and is called a 'punch-registration' system. It consists of a base plate, on which the film or paper is placed, and a set of cutting pins which are used to punch holes in this material. The medium employed has to be accurately positioned with respect to a 'side-bar' or 'end-stop' in order that the same location can be maintained for every sheet used. More sophisticated equipment allows variation in the adjustment of the end-stop, so facilitating the appropriate punching of sets of materials exhibiting a variety of formats. It is also possible to modify the number and character of the cutting pins used. The holes created in materials fit over register pins or 'studs' which are precisely machined to fill the available space without allowing any movement. Only in this way can an exact fit of two or more map elements be guaranteed.

Studs employed may exist as single items, or be manufactured as integral parts of a 'pin-bar', in which case their relative spacing cannot be modified. When using single studs, the greater the distance between two or more punch holes the more accurate the resultant registration (Fig. 5.19).

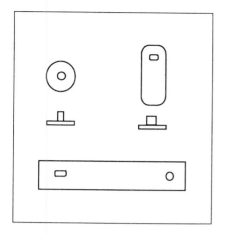

Fig. 5.19 Single register 'studs' and a 'pin bar'

There are a number of types of punch-registration equipment: that designed for single-side punching; devices permitting simultaneous punching on two or more sides; punches which create round holes; punches which cut slots; hand-operated systems; machinery powered by electricity or air pressure.

A number of the facilities mentioned are available in combination on some instruments (Fig. 5.20).

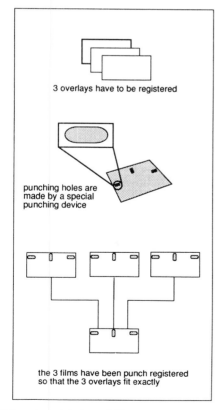

3 overlays have to be registered

punching holes are made by a special punching device

the 3 films have been punch registered so that the 3 overlays fit exactly

Fig. 5.20 Punch registering of films

Punch-registration can also be used in association with aluminium printing plates and thus make press preparation somewhat easier. In addition, its application helps cut down on labour and equipment costs. Punches used in conjunction with plates are of the 'heavy duty' type and are very strong and rather expensive.

The purchase of a good system should result in many years of optimum quality working. Once the equipment has been installed, maps and series of cartographic products should be produced to the same high standard of registration for as long as 20 years.

5.3.8.3 The use of registration

The precise registration of images, films, etc., is necessary throughout the pre-press and press-production stages. In traditional mapmaking, it is particularly applied to the following areas: process camera work (if possible or relevant); the imposition of material on a montage table; throughout contact photography and copy processing; the printing stage if more than one colour is to be applied.

Automated map generation also requires a type of registration, permitting the merging of base detail with new data and other information. In these cases no specific register marks are used, but co-ordinates or co-ordinate points serve as guides. If involved in the output of automatically produced maps or films from either a flatbed plotter or a raster-laser-drum-plotter, the use of register marks and/or punch-registration is strongly recommended.

5.3.9 Map component assembly

During map production the various components (lines, symbols, lettering/text and area tones) are normally produced on separate originals. Prior to the generation of plates, or colour proofs, these films have to be combined in order to produce one document for each of the intended colours. In order to understand fully the techniques involved, it is necessary to have read subsection 5.3.2 (in particular 5.3.2.1 Contact photography; 5.3.2.2 Films and papers; 5.3.2.3 Processing), subsection 5.3.6 (Platemaking for map printing by offset lithography) and subsection 5.3.8 (Image registration).

5.3.9.1 Film

The process will be described with the aid of simple flowcharts. In Fig. 5.21 only original negative materials are to be combined to produce a positive relating to detail intended to appear in blue (line-work, lettering, symbols and water areas). Precise registration can be maintained as a result of having used register marks and a punch-registration system throughout. In the case of new draughting foils or films, these must be punched before any contact photography takes place. Not all of the copying specified in the flowchart is made in contact (emulsion in direct contact with emulsion) as is the normal procedure. When exposing a negative through the base, the resultant positive will be slightly thicker and, conversely, a negative produced from a positive will exhibit some thinning (Fig. 5.22).

If during copying the individual, wrong-reading (WR) negatives are always in contact, assembly can take place directly onto a printing plate (Fig. 5.23). The same practice can be adopted for colour proofing.

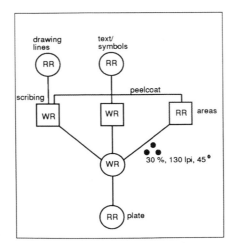

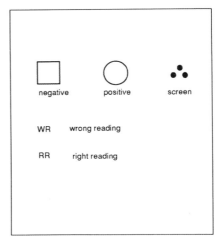

Fig. 5.21 The combination of original negative materials to produce a positive from which a printing plate can be made. NB: As yet no internationally accepted flow diagram symbolisation system has been devised

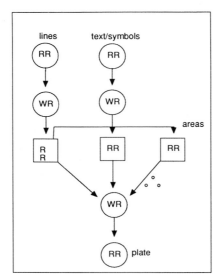

Fig. 5.22 Exposure and combination of right reading negatives to produce a wrong reading positive for use in manufacture of a right reading printing plate
NB: As yet no internationally accepted flow diagram symbolisation system has been devised

Fig. 5.23 Combination of wrong reading negatives directly onto a printing plate
NB: As yet no internationally accepted flow diagram symbolisation system has been devised

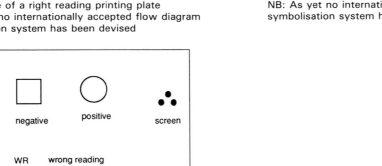

5.3.9.2 Montage for platemaking

The same punch-registration system used in the generation of final positives or negatives can also be employed during platemaking, and sometimes even on the printing press. This is why special pre-punched mounting foils are used. One is produced for each colour, and besides having punch holes also incorporates register marks outside the trim area. It is advisable to start by laying the 'black' foil, and subsequently to mount each of the other colours as separate overlays, each of which should be individually positioned on top of the first. These will each exhibit register marks duplicating those of the black which may also display trim marks relating to the intended final format. Colour control blocks will be positioned on each of the foils (normally at the bottom of the sheet and always outside the trim marks), and will be regularly checked with a densitometer during press production (Fig. 5.24).

5.3.9.3 Atlas page imposition

The laying of printing sections for atlas materials presents further problems. Besides all colours fitting together precisely, individual pages have to be 'imposed' in such a way that, after the sheet has been folded, they fall in the correct numerical sequence. A further complication arises from the fact that pages are normally required to 'back up' one another. Both sides of the paper will be printed on, and so special care must be taken to ensure that the imposition is correct and that overall registration is correct for the back, the front, and the back with the front (Fig. 5.25).

Folding: As far as possible, folds should run parallel to the grain of the paper, and always parallel with the spine of the intended volume. During printing the grain of a sheet should be parallel to the cylinder axis of the press (Fig. 5.26).

Other points for consideration during imposition: Remember to include register marks outside the limits of the intended format which should be demonstrated by incorporating trim marks. In order to economise on paper, plates and printing time it may be advisable to use half a sheet or employ a 'work and turn/tumble' technique. In this situation the stack of paper is manually turned once the first side has been printed, and work is then continued using the same plates. In both cases, obviously, the same colours are used on both sides of the sheet.

5.3.10 Reproduction by electronic means

Computers are playing an ever-increasingly important rôle in map reproduction. The modern graphic arts industries commonly employ scanners/plotters in their generation of a wide variety of illustrative materials. Modern photocopiers have taken over numerous straightforward tasks, whilst those which are more complicated and critical are performed by expensive devices which are starting to replace contact frames and process cameras. However, the technology and techniques used in cartography are so specific that the software used by graphic artists is not always appropriate. Currently it is possible to produce a map solely by automated means, but a number of problems have yet to be fully solved.

5.3.10.1 Scanning (basic principles)

For some applications the camera has been replaced by a 'scanner' which is used to measure the amount of light reflected by details comprising an original document such as a map. These reflection values are translated into binary notation representing the grey values, black, white or colours. Binary data can be sent to a computer for

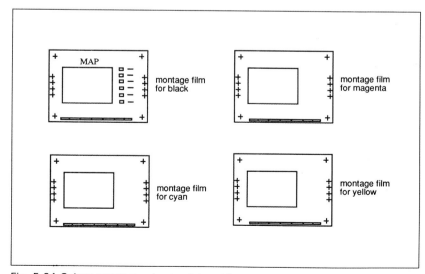

Fig. 5.24 Colour montage prior to platemaking

197

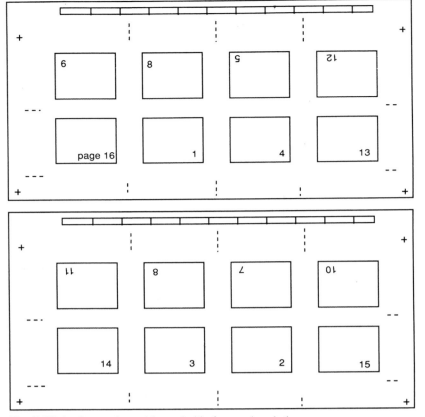

Fig. 5.25 An 8 page imposition suitable for use in printing
both sides of a sheet of paper

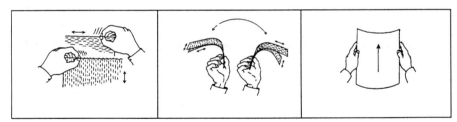

Fig. 5.26a Three methods of determining paper grain direction

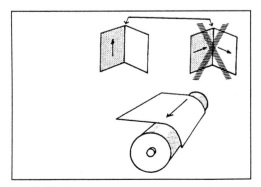

Fig. 5.26b The recommended paper grain for
printing and folding

storage and subsequent transmission to a work-station, or directly to an output device.

In the case in which plots are produced directly from scanned data, the computer merely transforms it into plotter instructions. However, in cartography the data, whether it be produced by scanning or digitising, must be able to be 'manipulated'. The reasons for this might be that: the quality of the data is poor and requires editing; data are not in an appropriate form and require combination; the data require graphic transformation by change of projection; scanned images in raster form need vectorisation; maps or map series may require revision and updating.

In cases of this type, data are sent to a 'graphics workstation' which provides a visual display on a screen. By using a cursor, mouse or keyboard the cartographer can manipulate or edit the graphic display to achieve the required standards. This is made possible by using software packages developed by companies specialising in the generation and supply of cartographic programs.

Scanned and manipulated data are stored in a specially reserved space on the disk termed the 'design file'. The latter allows only the inputting and modification of data, but no plotting is possible. If required, it will first be necessary to transfer design-filed detail into a 'plot file'. From this, output is possible onto materials such as paper, film or scribecoat, with the graphic being produced by a plotting pen, laser unit or scriber which moves from one co-ordinate point to the next.

Scanning analogue data: Scanners act, to a certain extent, like process cameras. The image is illuminated and then 'sensors' measure the amount of 'reflected light' emanating from it. The essential difference, as compared to a camera, is that this method of data capture stores it in a digital form within the computer 'memory'. Subsequently, the digital data can be modified by processing and manipulation. The manufacturers of scanners have developed a variety of ways for the viewing of information captured. Some of these present impressions of a scanned document as a series of lines, others as square or circular representations. The appearance depends on the type of sensors incorporated in the employed system. With respect to the line scanner, the larger the unit the greater the number of sensors required to provide high resolution. Low-resolution hardware incorporates a small quantity of sensors and the output is of poorer quality. Some instruments are capable of picking up the variations in light reflected from a complete map immediately, or at least information relating to significant areas. These are called matrix-CCD scanners (Charge Coupled Devices). The degree of illumination varies from type to type, but the end-products are comparable provided that colour sensing is not involved.

Two main types of scanners are used in carto-graphy:
—'Drum scanners' are the most accurate and dedicated sort. The image head scans whilst the drum carrying the map detail rotates. Sensors measure the amount of light reflected and data are stored in the computer. These instruments have the highest resolution facilities, and are therefore used for high-quality scanning (Fig. 5.27).

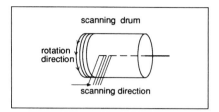

Fig. 5.27 A 'drum scanner'

—'Flatbed scanners' consist of a surface to which the original is held by vacuum. Normally the image head moves over the face of the map, but some equipment requires the document being positioned upside down on a glass plate. In this case the sensor bar, located beneath the document, traverses it from top to bottom (Fig. 5.28).

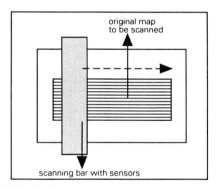

Fig. 5.28 A 'flatbed' scanner

'Scanners' are regularly used if large amounts of data are required from a significant number of maps. This situation particularly arises within bigger production agencies engaged in the preparation of numerous maps and design files for use in cartographic production. The scanning undertaken by mapping organisations is always followed by further data manipulation, so data is frequently stored on disk for some time prior to its interrogation and modification by an operator.

Colour separations: In order to be able to undertake 'colour separation', scanners must be equipped with a source of illumination covering the complete visible spectrum. Additionally, filters must be available for use during the process. This method is especially common in the graphic arts profession, but may be necessary in mapping when, for example, a colour photograph satellite image is available as a source of analogue data and colour selection is required. The original document, whether a multi-colour map or photograph, is separated to produce four positives or negatives. Each of these will contain its own specific analogue colour data, i.e. a different film for the representation of cyan, magenta and yellow elements of the original. A black film can be produced to enhance contrasts within colours and, if necessary, to produce genuinely black images (Fig. 5.29). Together, these four films serve to constitute a full-colour representation if processed through a colour-proofing system, or printed from plates in association with the appropriate inks. If this approach is followed, it may be unnecessary to manipulate or edit contained data.

The only things which might require adjustment are: the output contrast range; the output colour density; the possible rotation of screened output; the enlargement of the scale of the output; the area covered by the output.

If the scanner is a separate piece of equipment connected to the computer, manipulation and editing can be carried out prior to further production. Data will be stored in the computer and called up if it becomes necessary to employ a graphics workstation.

Proof scanning: If it is possible to produce a 'proof scan', this is strongly recommended because it provides a check, prior to formal/actual scanning, as to whether the size and/or scale is/are correct. Data resulting from a proof scan are stored in a kind of 'buffer' which can be erased after completion of the actual scanning. It should be stressed that manipulations such as rotation, skeletonising, attributing, colour changing, contrast variation, local density enhancement, etc., are performed after formal scanning is completed rather than during or after proof scanning. Application of this proofing procedure serves to prevent a lot of actual scanning needing to be repeated. Equipment time is expensive, and if alternative and more productive tasks can be carried out, this can render hardware use more profitable.

5.3.10.2 Data manipulation (applications)

This is carried out by using a graphics workstation which normally consists of a 'Terminal (VDU)', 'Keyboard', and a 'Digitising Surface' with a hand-held cursor (Fig. 5.30). Data are called for display on the terminal, and any required changes are implemented with the aid of the keyboard and cursor/mouse. The treatment given to 'data manipulation' in this section is deliberately brief, since it is explained in considerable depth in a variety of other textbooks.

As used here, the expression relates to the making of a significant number of corrections, together with changes of or additions to digital data, all of which are undertaken with the assistance of the computer. For example, if a text element is incorrectly positioned it can be moved by using appropriate manipulation software. The term 'data manipulation' also refers to factors such as change of projection, name placement, map layout, and even the positioning of photographic images or other illustrative detail.

—One of the most commonly required manipulations relates to the 'modification of tonal

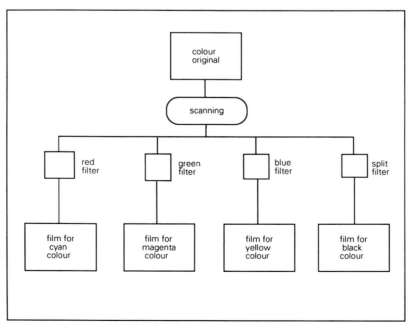

Fig. 5.29 Colour separation by scanning

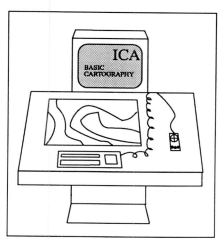
Fig. 5.30 A graphics workstation

gradations' and/or 'tone correction' of continuous-tone detail. The relevant part of the menu is activated using the cursor/mouse, and the system is asked to measure the included image densities. Results are displayed in the form of characteristic curves for each of the four colours comprising a multi-colour map, or for black and white in the monochrome situation. The operator then selects the detail he wishes to manipulate, and changes the characteristic curves interactively by using the cursor or keying in other values. As with traditional cartography, the selection of appropriate screen densities is still an important part of the effective display of detail!

— Apart from tonal corrections and gradations, the correct 'angling of screens' is a factor which must not be neglected. Halftone screens may consist of lines, dots or patterns. Any screen angle can be produced, but selection must be controlled by an awareness that visually disturbing moiré patterns can be formed when two screens are superimposed. In consequence, the well-known and documented precautions applied in conventional mapping should be adhered to.

— 'Masking' is another frequently applied function of the system, and is used to separate details, or parts of images, for further modification. It is a pre-manipulation function for possible use in tone correction and other data manipulation techniques.

— In some instances it may be necessary to retouch elements, an operation similar to adding shading or colour by spraying. Formerly it was essential to do this manually, but can now be accomplished using the 'retouch' and 'spray-retouch' functions within a program.

— The 'merging of continuous-tone images', such as aerial photographs, is an option infrequently used in cartography, but one which is regularly employed in the graphic arts environment.

— 'Pagemaking' options are included in even the most common programs, and relate to the process of assembling random elements such as text and other data to constitute a complete design. This facility offers a range of possibilities, and the more complex programs include a number of special options which are potentially useful in cartographic production. Thus the generation of borders, boxes, neatlines, register marks and other linework can be undertaken, although many software programs only allow this within the design part of a package.

Disciplines such as cartography are somewhat dependent on technical developments initiated within the wider graphics field. Consequently, specific tasks such as name-placement and generalisation have yet to be fully automated, because these are problems peculiar to cartography. Programs for name-placement are available, although requiring some modification, but those for generalisation have yet to be formulated.

It is for this reason that cartography is still in the process of transition from a traditionally based (analogue) to a modern, automated (digital) profession, and why this has happened rather more slowly than in the general graphic arts industry.

'Graphics workstations' allow their operators to visualise data captured by a scanner or digitiser, and act as a tool allowing the cartographer to see what has been done, what is being done, and what still remains to be done. This particularly relates to necessary modification, deletions, copying, etc.

Specifically developed cartographic software allows the operator to carry out necessary modifications and create a final design file which is stored on disk. The operating system enables conversion of design files to plot files via an applications option. If correctly compiled, a plot file controls operation of an output plotter. It is at this stage that one of the most common problems occurs, in that each type of plotter requires its own specific instructions and consequently the plot files are not always in a form which is immediately compatible with the requirements of a particular item of equipment.

5.3.10.3 Plotting

A design file cannot be used to generate hard-copy output on paper or film. Data must be transformed to an executable format in order to produce a hard-copy version of what was previously displayed on the screen of the graphics workstation. The 'plot file' must be available on disk where space is created to store the information required to make a plotter move, draw, engrave and/or expose.

'Plotters' are 'output devices', controlled by software, which create imagery reflecting design decisions made at the graphics workstation and

201

recorded in a plot file. There are a variety of different plotters on the market. In some instances the material receiving the image moves in the 'X' direction whilst the output device travels in 'Y'. Others allow base material to be mounted on a 'drum plotter' which rotates at high speed as a laser unit makes exposures whilst moving in the 'Y' direction. These are the most precise instruments, and are termed 'Raster-Laser-Plotters'.

Another group consists of the so-called 'Flatbed-Vector-Plotters', and involves output onto a large table which is used in association with a pen plotting, scribing or exposure facility to produce images on paper, scribecoat or film. Only linework is generated by this type of hardware, which does not create raster imagery (Fig. 5.31). From this description it should be evident that the most complete maps, incorporating screened areas, can only be produced via a raster-laser-plotter (Fig. 5.32).

Simple line maps can also be prepared using a less sophisticated vector 'pen-plotter', but the quality is often dubious. It depends on the preci-

sion of the pen and the overall accuracy of the equipment. This comparatively low cost hardware should only be used, in the ideal situation, for the production of check plots.

Recently developed techniques now allow reasonable quality output from 'colour plotters', and it is anticipated that, in a few years, excellent materials will be derived using this technology. Available plotters work in various ways, but it is not proposed to discuss these in detail here.

The most common type requires use of a film composed of cyan, magenta and yellow layers. Output onto this medium is by means of 'heated matrix pins' which can produce fine tones with a density of between 300 and 400 dots per inch. At present, the maximum plot size is A3, which is unfortunately often too small for cartographic purposes. However, use of these instruments guarantees very good colour appearance and predictable quality.

An alternative is provided by the 'ink-jet plotter' which utilises tiny insprays of cyan, magenta and yellow, the subtractive primary colours. The

Fig. 5.31 A flatbed-vector-plotter

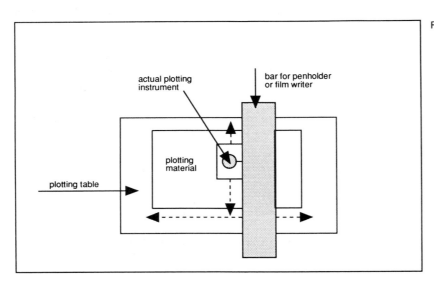

Fig. 5.32 A raster-laser drum plotter

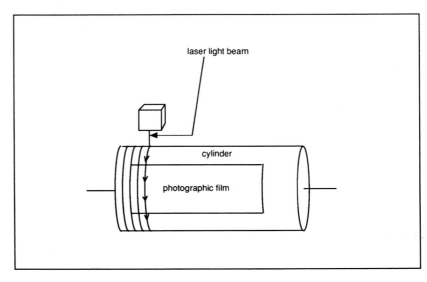

resultant output is of good quality, but the method tends to create stripes and has other inconvenient features. Consequently, this equipment cannot be used to produce a final product, although it is useful for the creation of colour proofs and 'discussion documents'.

High-precision film plotters are extremely expensive but generate output of excellent quality. In order to minimise investment, co-operation with other organisations could present an attractive option. However, if an arrangement of this type is entered into it must relate not only to equipment use, but also to the fact that all partners must prepare plot files in a form which can be read by the instrument purchased.

Less crucial accuracy requirements are reflected by lower hardware prices, but it should be remembered that cheaper output devices are only really suitable for the preparation of check plots. Production quality instruments are always very expensive. In today's graphic arts industry, many of the described techniques and technologies are well established, and colour reproduction is normally performed by means of scanners, graphics workstations and plotters.

If insufficient money is available for investment in this type of hardware, work can be produced through a bureau service. The final photographic films will be processed and returned either in a complete form or as small, page-size images. It is still essential to impose separate colour films as montages prior to in-house platemaking and eventual printing (see paragraphs 5.3.9.2 and 5.3.9.3).

5.3.10.4 Other methods of reproduction

Various new graphic reproduction techniques have recently become available and can, to some extent, be used in cartography. The most commonly employed method is photocopying, which is carried out by 'electrostatic copiers' of types regularly seen in shops and offices (Fig. 5.33). However, it is not this sort of equipment that is used in cartography, although 'colour copiers' have now become so sophisticated that they are occasionally used for map reproduction. Unfortunately, the output size from these instruments is normally rather limited, but larger formats are now being introduced and copy definition is constantly improving. It is also possible to modify colour strength without affecting the reproduction quality which remains consistent.

This represents a considerable advance, as formerly image density tended to decrease with increase in the number of copies produced. Official organisations involved in the printing of banknotes are somewhat concerned about these dramatic improvements in colour copying! Although it is feasible that this type of reproduction may replace lithographic printing for the generation of small numbers of copies, long runs will still be produced by using large offset presses.

If plotters can be classified under 'methods of reproduction', a number of extra developments can be cited here. An important differentiation must be made between:
  1. Raster-type plotters: ink-jet; bubble-jet; thermo-wax; dot-matrix; laser; for examples

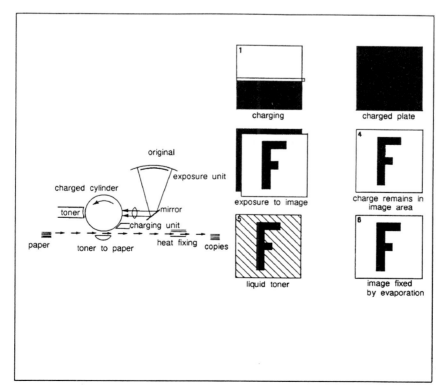

Fig. 5.33 Electrostatic copying

203

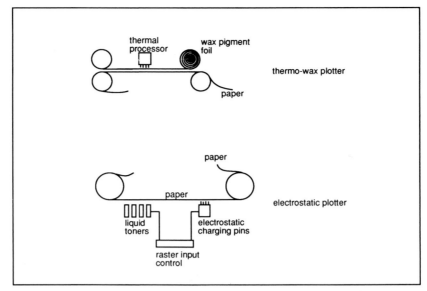

Fig. 5.34 Types of raster-plotters

see Fig. 5.34.

   2. Vector-type plotters: pen; scribing; cutting; photo-head.

This chapter would become too long if an attempt were made to describe the specifications and properties of all of these, so remarks have deliberately been restricted to a consideration of only the most commonly employed types.

A very new but potentially important development, which is likely to be used extensively for reproduction in the future, is the 'still video' technique. Essentially, this employs a camera which projects an image onto CCD sensors. These transform the electromagnetic radiation into digital data which are stored, in bits, on either a hard or floppy-disk. The cartographer calls up the data via a graphics workstation, and is then able to manipulate the digital map components.

At present, the system is still in an experimental stage, and some of its geometric aspects require improvement. However, it is expected that this method of reproduction will become commercially available in the near future, and that it will prove popular because of its relative cheapness and considerable capabilities.

5.4 Paper

This base medium is regularly used for a multitude of purposes; is produced in extremely large quantities; and in many types and qualities. It was expected that the introduction of the computer would result in a decrease in the amounts used, but quite the reverse has happened. Demand for paper products has increased significantly, and its manufacture is explained in Fig. 5.35. Figure 5.36 illustrates available paper sizes and their interrelationships.

5.5 Printing

The principal methods employed are letterpress, intaglio and lithography. Almost all map printing is carried out using offset lithography, which is an

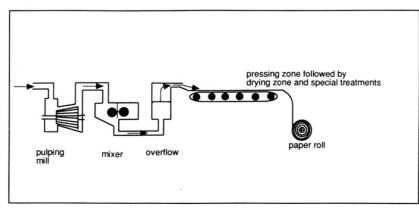

Fig. 5.35 Paper manufacture

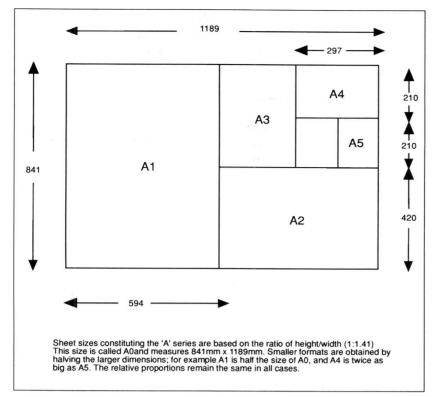

Sheet sizes constituting the 'A' series are based on the ratio of height/width (1:1.41) This size is called A0and measures 841mm x 1189mm. Smaller formats are obtained by halving the larger dimensions; for example A1 is half the size of A0, and A4 is twice as big as A5. The relative proportions remain the same in all cases.

Fig. 5.36 Paper sizes

'indirect' technique based on the principle that grease and water do not mix. 'Offset platemaking' is described in subsection 5.3.6, and the whole operation is best explained by means of an illustration (Fig. 5.37).

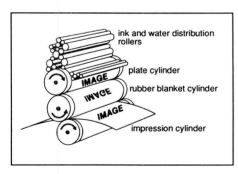

Fig. 5.37 Offset printing

5.5.1 Offset printing

A modern 'offset' press is capable of producing in excess of 10 000 impressions per hour, but in practice a running speed of between 5 000 and 10 000 copies is a more normal rate. Presses are manufactured to enable printing in one, two, four or six colours simultaneously. In order to speed up production, and render long runs more economical, manufacturers have developed equipment which can print on both sides of a sheet of paper at the same time. Reproduction is possible on paper at various sizes, ranging from A4 (210 × 297 mm)

to 2A (1 189 × 1 682 mm). The largest format is particularly intended for the simultaneous and accurately registered printing of four colours. Printed proofs and very short runs can be produced on a manually-operated, flat-bed press.

5.6 Finishing and presentation

Maps are always printed on paper which is larger than the intended end-product. Subsequently, the sheets have to be trimmed, and often folded, prior to presentation in their finished form.

5.6.1 Trimming

This process is performed by a mechanically operated guillotine which can be programmed for trimming to appropriate dimensions. The paper stack can be up to 10 cm thick, and has merely to be turned manually to expose its relevant edge to the blade. Trim marks are printed on the sheets to assist in the location of the paper.

The guillotine's blade is wedge-shaped and tends to move the stack slightly during operation (Fig. 5.38). Consequently, no two sheets have images in precisely the same position. This can result in problems of registration with maps extending over two pages in an atlas, but does not affect single sheets or those already trimmed and made subject to silk-screen overprinting, as is explained in 5.3.7.

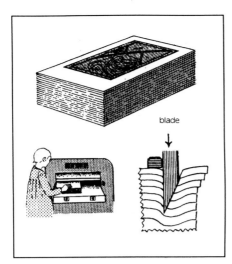

Fig. 5.38 Trimming by guillotine

5.6.2 Folding

This process is necessary to enable the easy handling of sheet maps and the eventual binding of the pages to be included within a bound atlas. For small editions (1 000 copies or less) the process may be performed manually, but for longer print runs specialist folding machinery is required. This can be adjusted to generate end-products exhibiting a variety of folded dimensions. There are two specific systems: knife- and buckle-folding (Fig. 5.39), which can be used independently or in combination. After appropriate setting of the necessary machinery, a folding rate of about 4 000 sheets per hour can be maintained. It is important to realise that the process is more straightforward, and that folds will be smoother,

if the paper is folded parallel to the grain. In the case of atlas maps, subsequently intended to be bound, this is vital and must be taken into account when planning the reproduction of a project.

5.6.3 Binding

Of necessity, paper sheets are bound when included within a book or atlas. This is, in itself, a complicated professional process and consequently is considered beyond the scope of this chapter.

5.6.4 Presentation

A map reproduced on a paper base is often flimsy, easily damaged or torn. Frequently, users wish to have cartographic detail presented in such a way that it is: in a more rigid and durable form; folded to exhibit an alternative format to that adopted by the publisher; suitable for exhibition as a wall display; usable as a working document which can be annotated or drawn upon. A number of possibilities have been developed to satisfy these requirements.

5.6.4.1 Plastification

The map document can be sandwiched between two plastic foils, which may be either clear or grained, and 'welded' along the four edges, thus rendering it both stiff and waterproof.

Alternatively, detail may be 'baked' onto a plastic surface, so making it highly durable and suited for outdoor use.

A third alternative is provided by 'decorative laminate' which consists of layers of special paper and thermoplastics which are bonded together, under pressure, to form a homogeneous material

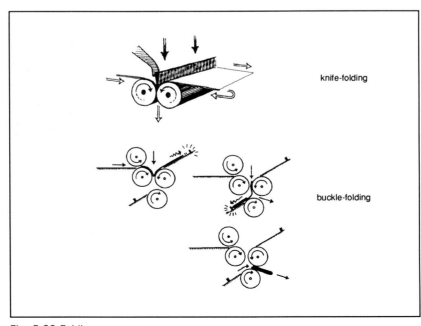

Fig. 5.39 Folding systems

with high resistance to both water and chemical attack (Fig. 5.40).

5.6.4.2 Lamination
During this process the map is covered by a tape-like, thin foil which may be either clear or grained. This can be stuck over one or both sides of the sheet and still allows the document to be folded (Fig. 5.41).

5.6.4.3 Varnishing
In order to enhance a printed surface's resistance to damp, and to make it water-resistant to a certain extent, it can be coated with a special varnish. This can be applied by means of a brush or aerosol spray. It should be noted that this does not increase the strength of the paper base. Maps which are intended for mounting should be printed using inks appropriate for the purpose.

5.6.5 Wallmaps

Not only maps intended for use in schools, but also a wide variety of other cartographic materials, may be displayed on walls. The most common types are:
- —maps mounted on soft-board, possibly laminated or varnished, and enabling the use of marker pins and tapes (Fig. 5.42);
- —portable wallmaps mounted on cloth and either with or without rigid rollers at the top and bottom. Modern cloth-backing materials have an adhesive layer onto which a paper map can be ironed. Subsequently, this map can be laminated or varnished. A further possibility is that wallmaps are printed directly onto synthetic paper (c. 200 g/m^2) which is then attached to rollers (Fig. 5.43).

Fig. 5.40 Using a water resistant map

Fig. 5.41 Using a laminated map

Fig. 5.42 A wall map mounted on soft-board

Fig. 5.43 A cloth-mounted wall map with rollers

INDEX

Figures in **bold** *denote figure citations*